P9-CCZ-466

MAKING DECISIONS

Second Edition

MAKING DECISIONS

Second Edition

D. V. LINDLEY
Formerly Professor of Statistics
University College London

JOHN WILEY & SONS
London · New York · Brisbane · Toronto · Singapore

Library of Congress Cataloging in Publication Data:

Lindley, D. V. (Dennis Victor), 1923–
Making decisions.

Includes index.
1. Statistical decision. I. Title.
QA279.4.L56 1985 519.5′42 85-12010
ISBN 0 471 90803 7 (cloth)
ISBN 0 471 90808 8 (paper)

British Library Cataloguing in Publication Data:

Lindley, D.V.
Making decisions.—2nd ed.
1. Probabilities 2. Decision-making
I. Title
519.2 QA273

ISBN 0 471 90803 7 (Cloth)
ISBN 0 471 90808 8 (Paper)

Printed and Bound in Great Britain

To Joan

PREFACE

to the First Edition

This book is about decision-making: about the logical processes that need to be used in arriving at a decision. It is not much concerned with the ways in which people currently make decisions. There is no material on delegating responsibility, on how the paper-work should be organized, or on the personality of the decision-maker. Instead we discuss the subject from a scientific viewpoint and see what basic principles there are in any choice of an action. We study the rules of decision-making. The book is addressed to business executives, soldiers, politicians, as well as scientists; to anyone who is interested in decision-making and is prepared to take the trouble to follow a reasoned argument. In particular, it is addressed to university students in *any* discipline.

The main conclusion is that there is essentially only one way to reach a decision sensibly. First, the uncertainties present in the situation must be quantified in terms of values called probabilities. Second, the various consequences of the courses of action must be similarly described in terms of utilities. Third, that decision must be taken which is expected—on the basis of the calculated probabilities—to give the greatest utility. The force of 'must', used in three places there, is simply that any deviation from the precepts is liable to lead the decision-maker into procedures which are demonstrably absurd—or as we shall say, incoherent.

The organization of the book is as follows. The first five chapters are devoted to a demonstration of the result stated in the last paragraph, together with extensive discussions of the twin concepts of probability and utility. It is here that the logical argument is tightest. It is included because no other discussion of the result can be so convincing as a demonstration of its inevitability. Chapter 6 studies how information can be used in decision-making and the next chapter discusses the value of the information. Decision trees, an important technical device for solving decision problems, are studied in Chapter 8, and a final chapter is a discussion of the conclusions reached in the study. There are several, mostly simple, exercises to which answers are provided.

In studying the logic of decision-making it is necessary to include a little mathematics. Every effort has been made to keep this to an absolute minimum. Most of the mathematical arguments that you will see on flipping over the pages can be avoided if necessary. There is, however, quite a lot of mathematical *notation*. This is essential for both brevity and clarity. Mathematical notation is only an addition to written language. Thus, if we wish to refer to the probability of striking oil at a particular spot after having had a favourable seismic test there, we could equally refer to $p(A \mid B)$ where $p(\ )$ means 'the probability of', A means 'striking oil at a particular spot', the vertical line $\mid$ means 'after having had' and B means 'a favourable seismic test there'. $p(A \mid B)$ is simply shorthand for the English phrase. The notation not only has the advantage of brevity but also makes it easier to understand the principles involved. We can concentrate on $p(\ )$ and A, say, without being confused with oil and seismic tests. We can see the wood without the trees.

There are many examples in the book. All are rather simple and perhaps trivial. But we cannot begin a study of flight by building a Concorde or a 747. Until the easy is properly understood, the difficult must remain out of our reach. What is reasonably certain is that the difficult will not involve any new principles. The technology of implementing the principles is yet to come. Appreciation, not execution, is our immediate aim.

It should be emphasized that none of the material in this book is new. I have not thought it right in a work of this character to cite the original sources. Many writers have been responsible for the notions developed here: principally F. P. Ramsey, B. de Finetti, J. von Neumann, L. J. Savage and most recently J. Pratt, H. Raiffa, and R. Schlaifer, in whose company I spent a most rewarding period at the Harvard Business School.

The stimulus for writing this book came from Leigh Edmondson who, by continually asking me to lecture on the ideas, encouraged me to think about the basic principles in an elementary way. A first draft was read by P. H. Jackson, M. R. Novick, N. Forward, K. C. Bowen, and D. G. Smith. I am most grateful to them for commenting on the material in such a helpful and constructive manner. But, above all, I must thank I. Wilson who read the draft with meticulous care. If the book has any grammatical virtues at all, it is due to him. Finally, it is a pleasure to thank the secretaries, N. Pateman, P. Rhodes, and M. Ware who typed, retyped, and calculated with such care, and my family who encouraged me at all times.

The quotations at the beginnings of the chapters are taken from the *Forsyte Chronicles* by John Galsworthy, with the permission of William Heinemann Ltd. I am also grateful to the Editor of the *Guardian* for permission to use quotations from that newspaper. A suitable text for the book is to be found in the *Guardian* obituary of Bertrand Russell. 'And if we catch ourselves thinking in moments of intellectual frigidity, "This sphere, at least, can never become the domain of cold reason", immediately we hear his not-very-ghostly voice asking ironically, "And why?"'.

London D. V. LINDLEY
June 1970

PREFACE
to the Second Edition

The first edition used a standard for probability, the fall of a point at random, that was artificial and unnecessarily complicated. This edition uses a standard of balls from an urn, which provides the basic results much more easily. It is still somewhat artificial and so a second method of practical relevance based on scoring rules has also been used. A major advantage is that realistic methods of probability assessment can be provided and an entirely new Chapter 9 is devoted to this and to the assessment of utilities. The former Chapter 9 becomes 10. It has become clear to me in the period between the two editions that the ideas of probability play a more important role in decision-making than I had previously appreciated. Consequently additional material on probability has been added and Chapters 2, 3, and 6 have been rewritten. The opportunity has been taken to make changes elsewhere, especially in Chapters 4 and 10. The text has been broken up into numbered sections in order to display separate topics to the reader and to make cross-references more precise.

Having much sympathy with most aims of the feminist movement, I have been puzzled as to what to do with the personal pronoun that, in its singular form, always refers to a sex. The form his/her is ugly. The plural, their, can sometimes be used. One can make the decision-maker feminine half the time, but this seems strained. So until there is a pleasant singular, his has continued to be used and I offer my apologies to any who feel uncomfortable with it.

2 Periton Lane D. V. LINDLEY
Minehead TA24 8AQ
England
August 1984

CONTENTS

Chapter 1

Decisions and Uncertain Events

'. . . all action is based on judgements . . .'

Flowering Wilderness, Ch. 14.

1.1 THE UBIQUITY OF DECISION-MAKING

The subject matter of this book is decision-making: a study of how decisions about which course of action to pursue *ought* to be made. All of us have to make decisions every day of our lives. Most of the choices involved are often trivial, as when a lady decides what dress to wear, or a diner selects from the menu: though one can easily imagine occasions when these are of moment. Every now and again we have to make decisions which we recognize as important and to which we devote a great deal of thought. Decisions concerning a proposal of marriage, accepting a new job, or buying a house will have important consequences that need to be carefully considered before coming to a conclusion. A few of us have to make decisions that affect our colleagues, and not just ourselves or our immediate families: managers in industry are such people, and the results of their decision-making will affect both shareholders and employees. An even more select group of people have to make decisions which determine the destiny of nations, and such statesmen usually occupy an extremely important position in society.

Decision-making is therefore something which concerns all of us, both as makers of the choice and as sufferers from the consequences, and there can be no doubt about the importance of the subject. It is therefore somewhat surprising that so little has been written about it from a scientific viewpoint. Historians and politicians have discussed in detail how particular decisions were made; lawyers have studied certain types of decision-making in great detail; but until recently the scientific eye had not looked at the field. The process had not been dissected in the laboratory; and literacy, rather than numeracy, had been the major talent of the investigator. Recently, however, some results have been obtained by statisticians which seem to be of importance to all concerned with decision-making (and therefore to everyone). The aim of this book is to inform a wider audience of these results, in the hope that they will find them of value in their work.

1.2 THE ROLE OF THE SCIENTIFIC METHOD

The contribution that the scientific attitude can make to decision-making merits some discussion. It is often held that the choice of a course of action is so much a human activity, dependent on the personality of the individual decision-maker, that the unemotional, abstract, and meticulous analysis of the mathematician is inappropriate. Napoleon did not need a computer to conquer Europe. The counter-argument points out that, although there is a strong human element in the process, there are parts of it that can benefit from some calculations. A leader of industry needs his accountants. The material in this book provides a tool to aid the decision-maker: it does not try to replace him. We would agree that the personality of the decision-maker is important in this activity, two aspects of this will be mentioned later in this chapter,* but there are parts that seem capable of systematic, analytical study. Specifically, the ways in which parts of a single decision problem can be combined to make the whole can be shown to obey certain rules. We do not offer a machine whose handle only needs to be turned to demonstrate the proper course of action: we merely provide some guide lines for sensible decision-making, guide lines which enable a complicated decision process to be broken down into smaller, and therefore simpler, parts whose separate analyses can be combined to provide a solution to the whole.

What the statistician has done is to stand back and look at the problem facing any decision-maker, and ask himself if there are any rules that must be satisfied before the decisions can be held to be sensible: or, to put it in a negative way, can we detect any patterns of behaviour that, if exposed, are seen to be ridiculous and must therefore be eliminated? In other words, the statistician has looked for consistencies that must be observed in any satisfactory decision process; and then he has gone on to deduce from the observed requirements of consistency, certain rules that need to be obeyed. The simplest way of ensuring consistency is to have a standard by which to judge any situation, and therefore much of the statistician's argument is concerned with the development of suitable standards.

1.3 COHERENCE

An analogy may assist the understanding, always provided one remembers that analogies are never perfect, and that the presence of some feature in the analogy does not imply its presence in the primary object of study. Consider a large and complicated building and concentrate entirely on the dimensions, forgetting that part is stone, part air; ignoring the uses of the different rooms, and the people who will inhabit them; thinking only of the sizes of the rooms, the doors, and the furniture. Then such a complicated structure has to have

* Imagination is needed in thinking of what actions are possible (section 1.6) and again in the contemplation of possible outcomes were a decision to be taken (section 1.9). In both cases, the decision-maker is trying to consider all possibilities.

certain rules of consistency in its dimensions. The windows may not be taller than the rooms, and must fit into their frames; the length of skirting-board plus widths of doors must add to the total wall-length of a room. The architect must take all such requirements into account in designing the building, and he does so by measuring the windows, ceiling heights and room sizes. The measurement is accomplished by referring them all to a standard, in Britain a foot (or a yard). Rather than compare directly the window with the ceiling height, he will refer both to the standard and express both in feet and inches. The fit can then be expressed by saying that the height of the window must be less than that of the room. What he has done is to express all the dimensions of the building in terms of a standard. We aim to proceed similarly with a complicated decision problem; by referring its separate components to a standard type of decision and using this standard to solve the major problem.

The analogy may be pursued a little further. The architect uses tools, sometimes quite elaborate ones, to measure the dimensions: for example, in the early stages he will use a theodolite in surveying the site. Similarly, a decision-maker will need tools. At the moment such tools are rather simple and are discussed in Chapter 9. Much needs to be done in the way of developing these, and, as a result, the reader will not find in this book all that he needs for the more involved decision problems. Nevertheless the basic ideas are here: the scientific principles are clear, the technological application of them is deficient. The business executive who looks for complete solutions to his marketing problems will be disappointed. To expect to have them is like hoping to find out how to build a bridge for a modern highway without having first learnt the laws of mechanics. Our aim is to provide the equivalent of these laws: the basic principles of decision-making. At the most we show how to build modest foot-bridges.

In the discussion just provided, the word 'consistent' has been used and we have talked about the requirement of consistency in decision-making. This is perhaps the natural English word to use but unfortunately it has other technical meanings in both logic and statistics. We therefore use instead the term 'coherent' and refer to coherence as a desirable property in choosing a decision. Essentially, this book is about *coherent decision-making*. The term will be defined precisely in succeeding chapters.

1.4 DESCRIPTION AND PRESCRIPTION

Before proceeding, mention must be made of one aspect of decision-making that will *not* be considered. We shall not study how decisions are made today in modern society. We suspect they are made badly: but this is not our concern except insofar as we hope to improve the situation by writing this book. To attempt such an empirical study would require extensive surveys and considerable laboratory work, including investigations of simple decision-making processes. Many studies of these types have been made and we shall have a little to say about some of them later in the book. They mostly appear to show that man does not make decisions in accord with the recipes developed here: in other words, he is incoherent. Such empirical results reinforce our belief that the statistician's contribution is significant: he appears to have something new to say; he is not confirming man in his present ways. Our method is not empirical; we sit back and think about the decision process, and show that it must have certain features. It might, unkindly, be called an ivory-towered approach. Perhaps it is, but come and join us in our ivory-tower for a while: I think you will find it leads to some results which are useful in practice.

A study of how people actually make decisions is called *descriptive*: its aim is to describe behaviour. Our approach, in contrast, is termed *prescriptive*: its object is to prescribe good methods that should be used in selecting a course of action. A third term, *normative*, is also used. The normative approach is a logical study of choice between decisions, strictly within a mathematical framework, and says how a person ought to behave: it provides a norm, or standard. For much of this book we shall be concerned with a normative analysis because it is this that provides the logical foundation upon which operational, usable methods can be based. This will be used to develop a prescriptive approach that attempts to blend the logic with the reality of the world. It will repeatedly be seen how the normative logic can be used to criticize contemporary decision-making and to demonstrate how it ought to be done. In particular, the methods developed here will be shown to be useful in the achievement of true democracy (section 10.5).

1.5 DECISION-MAKING AS A CHOICE BETWEEN ACTIONS

A person is faced with a decision problem whenever there is a choice between at least two courses of action. The lady who has only one dress does not have to decide which to wear. The first thing to do in any decision situation is to consider what courses of action are available. It is unnecessary to distinguish between decision and action. A decision to wear a particular dress need not be separated from the action of wearing the dress, for if the decision did not lead to the wearing of the dress it was presumably because something intervened to prevent it and a new decision problem arose. It is often inadequate to include one decision and its negation as a second decision, so giving a problem with just two courses of action. For example, a man is contemplating going to a theatre one particular evening: he cannot present this as a straight decision problem, to go or not to go to the theatre. For if he does not go he has to do something with his time that evening; to stay in the office and work, to go home and watch television, to read a book. There are really many courses of action that he might take that evening, and his decision problem is a choice amongst them, and is not a straightforward comparison of going to the theatre with not going.

This point is of wide applicability, and there are probably many important decision situations in which it is forgotten. I was once a member of a committee considering the erection of a new building. The lowest tender price was

£6000 more than the funds allocated to the building by another decision-making body. The chairman turned to the treasurer and asked if we could afford £6000 from reserves. He said, yes, and the committee agreed to the acceptance of the tender. This was a decision to spend £6000 on the building without considering alternative uses of the £6000 that might have been more beneficial to the organization. It is like deciding to go to the theatre without considering the alternatives.

Important decisions of government often appear to the ordinary citizen to be made without due regard to other possibilities, though it is one of the duties of the Opposition in a parliamentary democracy to draw the government's attention to them. A good example of this recently arose in Britain. A third airport was needed for London and the government set up a Committee to consider the suitability of one site, Stansted, without reference to other possibilities. The report produced aroused such opposition that eventually the government had to rethink the whole situation. I quote from a *Guardian* leader (13 December 1967): 'Lord Kennet was forced to admit that the planning procedure is unsatisfactory and is to be changed in future so that inquiries into major projects can consider the options instead of merely a "yes" or "no" to a particular scheme.'

1.6 AN EXHAUSTIVE LIST OF EXCLUSIVE DECISIONS

The first task in any decision problem is to draw up a list of the possible actions that are available. Considerable attention should be paid to the compilation of this list because the choices of action will be limited to those contained in it. It is sometimes possible to be certain that all reasonable alternatives have been included. For example, a retailer faced with the decision of how many items to order from his wholesaler could order any number between zero (that is, not order any) and the capacity of his store. These are the only immediate possibilities, though providing a bigger store might be another action to be contemplated. Often one cannot be sure that some attractive possibility has not been omitted: there is always the chance that some ingenious person will come along with a proposal that the decision-maker has not considered. It is almost certainly true that some successful decision-makers derive their success from their ability to think of new ideas, rather than from any ability to select amongst a list, so providing an example of the human element that was mentioned above. Such initiative and enterprise is to be encouraged and we can offer no scientific advice as to how it is to be developed. Our recipes will be limited to selection amongst a given set of decisions and it is important to make sure that the set reasonably exhausts the possibilities. When it does, it will be described as an *exhaustive* list. It is not possible to add on to a list another decision, 'do something else', without being more specific about what 'something else' is. This is because the notion is too vague to be used, like 'not going to the theatre'.

We therefore start with a list of the possible decisions that might be made. It is convenient to make it a requirement of such a list that only one of the decisions can be selected: that is, the possibility of choosing two or more is ruled out. For example, a menu is not such a list, because it is both possible and usual to decide on both an entrée and a dessert. The entrée part of the menu, however, is such a list because it is usual to decide on just one entrée. A list having the required property that only one member may be selected from it can easily be constructed from any given list by the simple expedient of taking all pairs, triplets, or larger sets from the first list and making them members of the new list. Thus a menu could be constructed where every item on it was a pair, one member of the pair being an entrée, the other a dessert. The decision about the meal would consist of the selection of a single item from this list, thereby providing both entrée and dessert. Such menus are occasionally found in Chinese restaurants. A list of this type will be longer than the original list but this does not cause any difficulty in the earlier stages of our argument. Later we shall see how to break the decisions in the earlier list into a sequence of decision problems, each with its own list. In the menu example the decision can be broken down into two parts: first, a choice of entrée; second, a choice of dessert. However it is easiest to begin with a single decision problem.

A list of decisions having the property described in the last paragraph is called a list of *exclusive* decisions. As explained in the last-but-one paragraph we also require the list to be *exhaustive*. Hence, the decisions are both exclusive and exhaustive: one of them *has* to be taken, and at most one of them *can* be taken. Alternatively, they *exhaust* the possibilities, and the choice of any one *excludes* the choice of any other.

The decisions in such a list can be written out in any reasonable order, and we can conveniently describe them in neutral language as decision number one, decision number two, and so on for as many decisions as there are in the list. (It is possible to imagine lists that are infinite and cannot be described in this way: but to obviate technical difficulties they will not be considered here.) It would be tedious to write and read 'decision number one' every time we wanted to refer to the first in our list, so we abbreviate the phrase in quotation marks to simply d_1. This is a shorthand which simplifies the description of the argument. Similarly we will have d_2, d_3, and so on. We want our discussion to refer to a list of any length and therefore we introduce a symbol for the number of items in the list. We use m, and therefore the last decision will be referred to as d_m. If there are five possible decisions (say five entrées on the menu) they will be called d_1, d_2, d_3, d_4, and d_5, with $m = 5$. For any value m the list will be written: $d_1, d_2, \ldots d_m$. The three dots, ..., are to be read as 'and so on up to'. Consequently $d_1, d_2, \ldots d_m$ is simply shorthand for 'decision number one, decision number two, and so on up to decision number m'. Mentioning the first two and the last one is purely conventional.

In summary: there is a list $d_1, d_2, \ldots d_m$ of m exclusive and exhaustive decisions and the problem is to make a selection, necessarily of a single item, from this list.

1.7 THE ROLE OF UNCERTAINTY

The selection of a single item from the list as being in some sense best and as the one course of action to adopt is, in principle, straightforward providing one has complete information. For example, the retailer cited above contemplating how many items to order from his wholesaler for immediate delivery would have no difficulty if he knew how many items he was going to sell before the next delivery. He would subtract from this number of future sales the number already in store and order the resulting quantity, assuming it did not exceed the capacity of his store, or zero if it were negative. The diner would have less trouble in selecting from the menu if he knew the quality of each individual item. ('I can recommend the trout today, sir: the manager caught them last night' is typical, valuable advice.) The difficulty in selection is usually due to uncertainties in the situation: to not knowing exactly what would happen if a particular course of action were to be adopted. For this reason the subject matter of this book is sometimes termed *decision-making under uncertainty*.

There are situations in which everything is known and yet decision is hard, but there the difficulty is purely technical. For example, consider any position in a game of chess. The legal possible moves can be listed and what will happen as a result of each one can, in principle, be described. The opponent will reply with his best move, you will reply with yours, and so on. This optimum play will lead to one of three possible results, win, draw, or loss. Any decision, or move, that leads to a win is best (provided such a decision exists). The reason we do not know what move to make is simply that the computations needed to find the best move are too complicated for even the largest and fastest of modern computers.

In this essay we do not consider such technical difficulties. It will be assumed that without uncertainty it is possible to say that one decision from the list is best, or that several are equally good and all these are better than the rest. It follows that all the decisions in the list can be ordered from best to worst, with possible ties. For we may consider a new list with the best one, or the best group, removed and choose the best from this reduced list. This choice will be second best in the original list. In order to select a course of action it will not, of course, be necessary to proceed with such an ordering but it can, if needed, be provided.

1.8 UNCERTAIN EVENTS

If the real difficulty in decision-making resides in the uncertainties in the situation, then these uncertainties have got to be considered in detail and introduced into the study along with the decisions themselves. Let us return to the retailer (section 1.6) wondering how much to order for immediate delivery. Suppose m denotes the size of his store, so that if his store holds 12 items with no room for any more, then $m = 12$. Assume, for simplicity, that at the moment when he is deciding how many to order, he has none in the store. When he writes an order he has available m exclusive and exhaustive decisions d_1, d_2, ... d_m; d_1 meaning the decision to order one item, and the other decisions similarly. If he orders more than m he will not have room in his store. Let us denote by j the number of items he sells before the delivery of the order after the one he is now considering making. It will be assumed that j does not exceed m, since he has, at most, m items to sell. Also we suppose j is at least one; if j were zero and he knew this to be so, the retailer would not place an order, a case that has been excluded. It is clear that if he knew the value of j he would order exactly j items, that is, select d_j. For then he would have no unsatisfied orders and again have an empty store with no items remaining unsold. Unfortunately the retailer does not usually know j, the number of items he will sell. He may have some idea of the value based on previous experience: for example, he might feel fairly sure of selling 2 but think it unlikely to sell as many as 25, but typically he will not know j. It is this uncertainty that we have to capture in our analysis of the situation.

The uncertainty in any decision-making situation may usually be expressed by saying that we do not know what will happen on some future occasion, though sometimes the uncertainty is in the past. We use the term *event* to refer to an occurrence or happening. Thus we speak of the event that the retailer receives four orders; the event that the market for a product is a million a year; the event that the trout is good; and so on. An event can either be known to have occurred or not, and it may change its status: thus the retailer will know the number of orders eventually, though at the time of placing the order he was ignorant. An event about which we are informed as to whether it did (or will) occur or not is called a *certain* event. Otherwise it will be referred to as an *uncertain* event. Thus the event that the retailer's store is empty is certain, the event of selling four items is uncertain. Notice that the word 'certain' has at least two meanings: 'particular' and 'sure'. Here it is being used in the latter sense. 'Uncertain' fortunately has no such ambiguity.

We denote by θ_j the event that the retailer sells exactly j items. Here θ is the Greek* letter, theta (pronounced to rhyme with heater or hater according to which side of the Atlantic you favour, the vowel being longer in the east) and θ_j is simply shorthand for the part of the first sentence which follows it. Then we have m such events θ_1, θ_2, ... θ_m. If the retailer knew which event was going to occur, his decision problem would be simple. But he does not have this piece of information, so the events are uncertain. Notice that, like the decisions, the uncertain events in our example of the retailer are exclusive and exhaustive: exclusive because one of them occurring excludes the possibility of any other taking place; exhaustive because one of them, by supposition, must take place. It is possible to ensure exhaustion by adding an event θ_{m+1}, 'something else happens', but this is usually too vague to be useful, as with the introduction of a vague decision to do something else.

* It may seem unnecessary to invoke another alphabet but experience has shown that it is very convenient to separate the uncertain elements in the situation from the known ones (like d_i) by using the Greek alphabet for one and the Roman for the other.

1.9 AN EXHAUSTIVE LIST OF EVENTS

A second example in which the uncertain events are not so easy to enumerate will serve to carry the argument a little further. A man lives by the sea at one side of a mountain chain. He has an appointment in a town on the other side of the mountains: the time is midwinter. There are only two ways to travel: he can either drive his own car over the mountain pass, or he can take the train. There are therefore only two decisions: d_1, to go by car; d_2 to take the train. Uncertainty enters in because he does not know whether the pass is blocked by snow. The weather high up in the mountains is very different from that down on the coast, so it is almost useless to look out of the window: weather forecasts are more concerned with populous areas than mountain-tops, so are not informative. He knows from past experience that the motoring organizations are always pessimistic, acting on the principle that it is best to keep the motorist off the road so that they won't have to rescue him should he get stuck in a snowdrift. What is he to do?

A simple analysis suggests that there are two uncertain events: θ_1, the pass is blocked; θ_2, the pass is open. These are clearly exclusive and exhaustive. But some reflection convinces the motorist that this does not adequately reflect the uncertainty relevant to the situation. For suppose the pass is open but the road is icy and the visibility poor. Then he will have to drive slowly and carefully and he will arrive at the appointment tired and ill-tempered. On the other hand, if the pass is open and the weather sunny he will derive considerable pleasure from the mountain scenery and arrive exhilarated and refreshed. This suggests three uncertain events: θ_1, the pass is blocked; θ_2, the pass is open but the weather bad; θ_3, the pass is open and the weather good. These are still exclusive and exhaustive events. But further contemplation may suggest to him the possibility of having an accident in the car, sufficiently serious to prevent him making the meeting in time. And if an accident in the car why not one in the train? No, he decides, that possibility is too remote to consider; but it does remined him that he might miss the meeting through the train being late—and that is a real possibility. These considerations suggest introducing yet more uncertain events: we do not list them all, but typical members of the list are 'pass blocked, train late, or 'pass open, weather bad, accident to car'. The reader might like to try providing himself with a list. He should arrange it so that the uncertain events are exclusive and exhaustive. This can always be done in a manner like that used to make the items in the menu exclusive in the earlier discussion of decisions. An important point to make with this example is that the uncertain events should cover all the contingencies likely to affect the decision-making, such as accidents to the car or delays to the train. On the other hand, irrelevant events such as whether the tide is ebbing or flowing can be omitted: they are irrelevant because they would not influence the choice of car or train. This example will be discussed in detail in section 8.12.

As a result of deliberations like those in the example, it should be possible to provide a list of uncertain events that cover the contingencies likely to affect the choice of a decision. Furthermore these uncertain events can be made exclusive and exhaustive, so that one, and only one, of them will occur. As with the list of decisions, there is always scope for ingenuity in altering the uncertain events. Someone may think of a relevant factor that the rest had forgotten. The ability to judge the relevance of factors is one of the skills in decision-making. The number of the uncertain events will be denoted by n, and the events by $\theta_1, \theta_2, \ldots \theta_n$. The retailer example was special in having the numbers of decisions and events equal, $m = n$.

In summary: there is a list $d_1, d_2, \ldots d_m$ of m exclusive and exhaustive decisions; and a second list $\theta_1, \theta_2, \ldots \theta_n$ of n exclusive and exhaustive uncertain events. The problem is to select a single item from the first list, not knowing which member of the second list will be true.

1.10 THE MODEL FOR DECISION-MAKING

This is the basic model we shall use in our discussion of decision-making. It may well be argued that the model is too simple, in that it ignores certain features of a typical decision situation: or alternatively, that it is too complicated, in that it is often impossible to consider all the decisions or all the uncertain events. It is always difficult to judge whether a model is adequate or not. For the purpose of predicting the motions of the planets it is possible to use a model in which the planets and the sun are solid spheres: ignoring the facts that the sun is hot, the earth is cold, Venus has clouds, or Saturn has rings. These facts are irrelevant to considerations of planetary motions. We feel that our model contains all the details relevant to decision-making: but the only argument we can adduce is to ask the reader to pursue it and see where it leads. The proof of the pudding is in the eating. Of course, there are tremendous technical difficulties in providing the two lists in some cases. But, as in our analogy with length, discussed above, these technical points can probably be overcome.

In any case, I suggest that much light can be thrown on a complicated decision problem by the mere attempt to provide these two lists. The realization that decision-making involves a comparison of alternatives, so making it necessary to consider them, is, in itself, an advance. The realization that it is necessary to consider the uncertainties that may affect the situation makes people contemplate the outcomes of their decision-making much more carefully. I go so far as to suggest that even if the rest of this book were ignored and only the ideas of this chapter, leading to the model, were used, then decision-making would be greatly improved. When considering the possibility of building a nuclear power station are alternative energy sources always considered? Is the possible event of not having enough electricity available to use domestic appliances adequately appreciated? Do advocates of increased nuclear armament spell out the events that their employment might produce? The hard contemplation of the lists $d_1, d_2, \ldots d_m$ and $\theta_1, \theta_2, \ldots \theta_n$ can be enormously clarifying. The reader may like to take up today's newspaper, consider one of the many examples of decision-making reported therein, and subject it to the analysis of this chapter. The insight it provides will usually be worthwhile, though this may reflect on the journalism, that does not report in sufficient detail, as on the decision-makers.

1.11 SUMMARY OF THE BOOK

In the next chapter the uncertain events are discussed. It is shown how the uncertainty can be described numerically in terms of probability. Since this point is so important, two justifications are given: one in terms of a standard, another employing the device of scoring rules. The former method is conceptually simpler, the latter is more useful. Probability, it turns out, is basic to all action under uncertainty. Probability obeys laws and these are discussed in Chapter 3. These laws are the essential devices that secure coherence (section 1.3) and are illustrated by several examples.

A course of action leads to a consequence, and the fourth chapter develops a second numerical scale to describe the merit of consequences: desirable outcomes having high values, undesirable ones, low values. This measure of merit is called utility and Chapter 5 parallels Chapter 3 in discussing the coherence requirements for utility similar to the way the earlier chapter discussed probability. The two numerical concepts of probability and utility combine to form an expected utility and the main result of this book is simply expressed by saying that decisions should be made by maximizing expected utility.

The natural reaction of anyone having to make a decision under uncertainty is to remove as much of the uncertainty as possible by acquiring more information. Chapter 6 describes how the additional information can be incorporated into the analysis. The basic result is a famous, or even notorious, theorem due to Bayes, an eighteenth-century clergyman. Even if information is to be had it typically costs something to obtain it and Chapter 7 is concerned with the value of information, or simply how much one should pay for it.

The reader has already been warned that the main topic of this book is the basic structure of decision-making and that less attention is paid to current important but complicated decision problems, partly because the necessary technical apparatus is unavailable. But one important technical tool is the decision tree. Chapter 8 describes the idea and illustrates its use with several examples. Again, paying attention to practicality, the ideas in this book are of no use unless the required measurements of probability and utility can be made. Methods whereby these twin tasks can be accomplished are described in Chapter 9. In the final chapter some deficiencies of the methods developed in the book are described, as are some more speculative extensions to conflict situations and important issues like democracy and freedom of information.

Exercises

1.1. A lunch menu contains only two items: a meat dish and a fish dish. You think the meat can either be good or poor; the same with the fish. List the uncertain exclusive events and the decisions. Further contemplation suggests another decision: go and eat elsewhere. Yet further contemplation suggests that another relevant event is what your wife is going to give you for dinner at night: suppose again that this is only meat or fish. Repeat the exercise. Suppose you decide that provided the meat is good you will settle for it, irrespective of the state of the fish or your wife's choice of food. Show that not more than 5 uncertain events are needed.

1.2. A transport manager of a firm has to order some new lorries. There are only two sorts available, 5-ton or 10-ton, and he requires 30 tons capacity. He therefore says there are two decisions: d_1, buy 6 of the lighter lorries; d_2, buy 3 of the heavier. Is this list reasonably exhaustive? Provide an exhaustive list of decisions.

1.3. A production line can produce units of three types A, B, and C. A day's work contains three shifts. All the units produced in a single shift are of the same type. The production manager has the task of saying at the beginning of each day what units shall be produced in each shift. If only the *total* production for a day matters and not whether a type was produced on the first or any other shift, list his exhaustive and exclusive decisions. If the demand on any one day for each of the types is either zero or equal to the product of two shifts, list the possible uncertain events in the situation. If at the beginning of the day, the stock contains no units of type A, one shift's worth of type B, and two shifts' worth of type C, consider the merit of some of the decisions available to the manager.

1.4. A patient has one of four illnesses, θ_1, θ_2, θ_3, θ_4. The doctor has three treatments available, t_1, t_2, t_3. He can apply any treatment and, if it fails, apply another, and if both fail, apply a third. The effectiveness is described in the following table

	θ_1	θ_2	θ_3	θ_4
t_1	1	0	1	1
t_2	0	1	0	1
t_3	0	1	1	0

where a 1 means effective and 0 ineffective. Enumerate the possible sequences of treatments (that is, the decisions) the doctor might take.

Chapter 2

A Numerical Measure for Uncertainty

'If one thing is more certain than another—which is extremely doubtful—'

Maid in Waiting, Ch. 13.

2.1 SOME EVENTS ARE MORE LIKELY THAN OTHERS

In the previous chapter a decision problem was broken down into two lists; of decisions d_i and of uncertain events θ_j. The next two chapters are concerned entirely with uncertain events, not just those that arise in a specific decision problem but any events whose outcome is unknown to you. Examples abound: the mountain pass is blocked (section 1.9), it will rain tomorrow, the retailer will receive 10 orders in the next period (section 1.8), Shakespeare wrote the plays attributed to him. There is no shortage of examples, for we are surrounded by uncertainty. Although events may be uncertain, in the sense that it is not known whether they are true or false, whether they will occur or not, some events are more likely to occur than others. It is more likely that a newly minted coin will fall heads when tossed fairly than that you will die this year. In the example of the retailer, cited in the previous chapter (section 1.8), where the uncertain events corresponded to the number of orders he would receive in a future period, he might well expect to have about 10 orders, say, but to have as few as 2 or as many as 25 would be distinctly unusual. The English language has several words to describe aspects of the uncertainty that is felt about events: *likely*, *probable*, *credible*, *plausible*, *possible*, *chance*, *odds*, and many others. The richness of the language reflects the ubiquity of the concept of uncertainty. All these words express the idea that some events are more likely (or whatever the word is) than others. Now the easiest and most useful way to order things is by means of numbers, so it is natural to use this device in discussing uncertain events. Our aim is to describe the concept of uncertainty numerically: for number is the essence of the scientific method and it is by measuring things that we know them. Specifically what we want to do is to attach to any uncertain event a number that describes that uncertainty.

We shall arrange it so that the larger the number the more likely the event is to occur: thus the event of 10 orders will have a larger value than either 2 or 25 orders, perhaps 0.2 for the former and only 0.01 for either of the last two.

Many people will immediately object that it is ridiculous to suppose that *all* uncertain events can have their uncertainties measured; some are so complicated, so involved with emotional issues, that it is absurd to reduce this intricacy to anything as simple as one number. In reply, a scientist might argue that measurement, through numbers, has proved such a useful device that it might be worth aiming for a numerical description of uncertainty for to be successful would open up many possibilities. Furthermore, it can be done, as will be demonstrated in this chapter. The difficulty lies with the actual measurement, not with the concept. As with other numerical values, like length, the idea is simple, the actual measurement may be hard. It is the idea that here concerns us. So let us try the measurement game and see what happens.

2.2 EVENTS WITH A NUMERICAL DESCRIPTION OF UNCERTAINTY

First notice that there are some uncertain events that are already described numerically. A class that immediately springs to mind is of events connected with horse-racing. A horse will be quoted at odds of 5–1, say, the larger the odds the less likely it is that it will win. The odds, 5–1, or simply 5, provide a number measuring the uncertainty of victory for a particular horse in a particular race. Another class of events is encountered with games of chance. These games incorporate events that are deliberately unpredictable, like the fall of a die when thrown or the drawing of cards from a well-shuffled pack. For example, with a die we might say that each of the six faces was equally likely to be uppermost when the die is rolled or that the chance of a six was 1/6. Similarly, the chance of an ace from a well-shuffled pack of cards is 1/13. Notice that the bookmaker uses odds; games ordinarily use chances. However, the two measures are connected; we could say the odds against an ace are 12 to 1, or the chance of a horse at 5–1 winning is 1/6. The precise connection between chance and odds is discussed in section 3.11.

There are other ways of measuring the uncertainty associated with events. Demographers keep statistics on the numbers and ages of populations and from these data actuaries are able to calculate expectations of future life. Consider yourself: it is uncertain when you will die but by reference to actuarial tables you could find your life-expectancy based on the group of people to which you belong, for example, white, male Britons aged 60. These figures are treated very seriously by insurance companies who effectively bet on your dying, though they do not use the language of the bookmakers. The conditions of the insurance, the premium, the payment when death occurs, etc. all depend on the actuarial figures for life-expectancy. The measurement of uncertainty is central to the whole activity of life insurance. But insurance does not only refer to life, it embraces many other forms of human activity: indeed, it has been said that one can insure against anything. Suppose that you possess a valuable object like a camera and you are concerned that you might lose it. Then it is possible to take out an insurance to cover it against loss for a period, perhaps a year. The premium that will be charged for the policy will depend not only on the value of the camera but also on the chance that it will be lost (section 2.13). Thus the premium when you travel abroad with the camera will exceed that required if you are to keep it in your house. The whole activity of insurance is concerned with uncertain events like loss of a camera, death or damage to one's health, and the measurement through money that the insurance companies use is partly a measurement of the uncertainty of events.

2.3 STATISTICAL AND NON-STATISTICAL EVENTS

It is undoubtably true that *some* uncertain events have their uncertainties measured: the question is, can the numerical assessment be made for *all* uncertain events? If insurance is accepted as a measure, and if it is true that anything can be insured, then the answer is yes. We might say that Lloyds of London is the measuring device. However, insurance is not a very good measuring instrument because other considerations are involved in the transaction besides uncertainty. Money obviously, but, more importantly, the perception of the value of money as perceived both by the insurer and the client. We shall study this later (section 5.12) under the name of utility, but for the moment we need only note that the insurance device is rather complicated, though we return to it in section 2.13. If this is dismissed, then the question of a measurement for *all* uncertain events remains open.

One possibility is that some uncertain events can be measured, like those in games of chance, and others cannot, like the event that Shakespeare wrote the plays ordinarily attributed to him. Now one thing that distinguishes these two events is that the former can be repeated almost indefinitely whereas the latter cannot. (Actually no event can be exactly repeated—at least the time will have changed for the second event—but some events can be repeated with the relevant conditions remaining unchanged. Often time is irrelevant.) The coin can be tossed, the die rolled repeatedly but there is no repetition of the Shakesperian plays being written. There is an element of repetition with the bookmaker because he is concerned with a whole series of races; and with the insurer who has a large number of policies; but not so much with the punter or the camera-owner. A distinction is often drawn between events which are *statistical* and those which are not. Statistical events are those which are capable of extensive repetition under essentially similar conditions. Non-statistical events are essentially unique. Games of chance are obviously statistical since they can be played over and over again. From the actuary's point of view, death is also a statistical event, since he considers a large population at risk and the event occurs whenever a member of the population dies. On the other hand, William Shakespeare is a unique person and the plays are a unique piece of writing; it would be ridiculous to put him or them in any group and use a statistical argument over the disputed authorship.

2.4 DIFFICULTIES WITH THE SEPARATION OF EVENTS INTO STATISTICAL AND NON-STATISTICAL

The idea of dividing uncertain events into statistical and non-statistical ones, and attaching chances or probabilities to the former but not the latter, therefore looks promising. Several people have tried to do this and some will only admit the possibility of probability arguments in the repetitive cases. There are three snags to the attempted dichotomy: first, on close inspection it becomes quite hard to decide whether some events are statistical or not; second, purely statistical probabilities are useless in many situations; third, it is sometimes ambiguous just what repetition means. To illustrate the first snag consider the example of a driver contemplating driving over the mountains and the event that the pass will be blocked by snow. Is the event statistical or not? The meteorological records for the district over a period of years could be obtained and a statistical probability for blockage could be calculated, as with London's rain in section 2.7. However, suppose that the local authority has recently improved the road and bought new snow-clearing equipment. Is the statistical probability the correct one? Clearly it is not and yet the event has not changed. So is the event statistical at all? The local authority's action has not changed the event in any way: it has altered the relevance of climatological data to the event.

The second snag of the uselessness of statistical probabilities may be illustrated by considering perhaps the most important uncertain event of our age: the event of a major nuclear war before, say, the year 2000. Repetition is clearly an irrelevant and ridiculous notion. Uselessness, even in a statistical context, may be illustrated by the actuarial example of death. To the actuary contemplating a population the event appears undoubtably statistical but to the individual the statistical relevance is not obvious, because he is concerned with his own death and not with any population of people. He is an individual: the actuary looks at the masses. The difficulty even arises for the actuary who has to make a choice of which population the individual belongs. Consider the actuary assigning myself to a population. He will surely take account of the fact that I am British and only consider the white population of these islands. Also he will take account of sex and age. So there is a population of white, British males aged 60. Fifty years ago he might have divided the population into teetotallers and not, and assigned me to the latter. Today he is more likely to take account of smoking habits and reduce the group to non-smokers. But will he take into account the ages of my parents' deaths, the fact that I am married, or many other features? If he does, the population will be reduced eventually to one, myself, and I will be seen to be as unique as Shakespeare.

The third snag of ambiguity in what is to be repeated may be illustrated by a person who tossed a coin three times obtaining heads, for the first time, on the last toss. What should be repeated: sets of three tosses, or sets of tosses ending with the first head? You might not feel it matters, but it does and the snag lies at the root of much statistical controversy.

The contrast between the long-run behaviour and the unique event occurs even in games of chance or at the race-track. The casino operator or the bookmaker is involved with the long run and expects to lose some days, but the player or the punter is much more concerned with his single visit or even with a single turn of the roulette wheel. We shall find it convenient later in this book to distinguish between the long-run behaviour, the statistical value, and the individual occasion. We shall use *chance* in connection with the former and probability with the latter. Our approach will begin with the unique event and only later incorporate the repetitive element when it exists. For example, consider first a single toss of a coin and the probability that it will fall heads; then contrast this with the frequency, or chance, of heads in a long series of similar tosses. The great advantage of the approach using probability based on the uniqueness of the event is that it is always applicable whereas the frequency notion is not (as with Shakespeare). Furthermore, the decision-maker is almost always concerned with a particular event that arises with a single decision that has to be taken, not with a sequence of decisions. Therefore the probabilistic approach is the practically useful one.

To summarize the argument so far; we aim to measure the uncertainty of any event, some events have their uncertainties measured statistically but this measurement is not always available and, even when it is, is not always the relevant one. Let us therefore ignore the statistical approach and consider an alternative one.

2.5 MEASUREMENT BY REFERENCE TO A STANDARD

Any measurement is constructed by reference to a standard. Length is described in terms of the wavelength of sodium light; time by reference to the oscillation of a crystal. It is therefore sensible to attempt the same comparative technique when measuring uncertainty. Before doing this note that actual measurements are not made by using the standard. We do not assess the size of the table by sodium light; a tape-measure or similar device is used. Consequently the reference to a standard for uncertainty is not usually a practical way of measuring it. Rather it provides a definition and, more importantly, enables important properties of the measure to be found. A vital feature of numerical uncertainty is the rules that it has to obey.

To provide a standard consider a container, traditionally called an urn, whose contents cannot be seen but are easily removed. The contents are 100 balls as near identical as possible except that some are coloured black and the rest white.* A ball is drawn from the urn in such a way that you think each of the 100 balls has the same chance of being drawn. (This phrase will be made precise in a moment.) Consider the uncertain event B that the withdrawn ball is black. The uncertainty clearly depends on how many black balls are truly

* In the United States the classical urn is sometimes replaced by a book-bag, the balls by poker chips.

in the urn. If b are black, and 100-b white, the probability of the event B is defined to be $b/100$ or $b\%$. Thus, if 50 are black, the probability is $\frac{1}{2}$ or 50%. This is the standard to which all uncertain events will be referred: or rather, the set of standards for differing numbers b of black balls from 0 to 100.

Now consider any uncertain event E. To fix ideas take the event that it will rain tomorrow in London. Now suppose you were to be offered a small prize if the event occurred: if it did not, you would get nothing. No stake is involved. Next, suppose you were to be offered the same prize if a black ball were to be drawn from the urn under the conditions already described. That is, there are two gambles, one contingent on E, rain, the other on B, a black ball, but otherwise identical. Granted that you may only have one gamble, which do you prefer? Again it depends on the number b of black balls. If there are none it would be best to gamble on rain: at the other extreme with all black balls, the urn is better. Generally, the more black balls the better is the urn gamble. It easily follows that there must be a particular number of black balls such that you are indifferent between the two gambles: call this number b. Were there $(b+1)$ balls the urn gamble would improve and be better than the rain one: with $(b-1)$ it would be worse. The event B has probability $b/100$ or $b\%$. Since the two gambles are now in all respects equivalent we say the probability of E, rain tomorrow in London, is also $b\%$.

An objection may be raised at this point. In considering the withdrawal of a ball from the urn the value of $b/100$ is firm and sensible: with b black and 100-b white balls and the stated conditions, the value is clear. But in considering the event E, the value $b/100$ is imprecise and, if not ridiculous, certainly inadequate: b could easily be changed by 1 or 2. In other words, the urn probability is clear, the event probability is vague. Perhaps, as several people have suggested, we need a second number to describe the assurance that the probability has; high for urns, low for E. In fact, as will be seen in section 6.13, our approach using a single value incorporates a measure of assurance. Indeed one of the best justifications for the arguments in this book is that they contain methods for handling objections. It is a justification that cannot be given until the calculations dependent on the assumptions have been made but is impressive because of the revelation that the calculations give of its hidden merits.

In this way any uncertain event E has its uncertainty measured by a probability $b/100$. The value b is just the number of black balls that would make you indifferent between the two gambles.

2.6 SOME COMMENTS ON THE STANDARD

In constructing the standard it was supposed that 'each of the 100 balls has the same chance of being drawn'; an essential restriction because if you felt that a black ball had a greater chance than a white one your chosen value of b would be affected. Since this requirement employs the word 'chance', clearly related to probability that we aim to define, some have held that the definition is circular. This is not so. The requirement in quotes can easily be defined and the circularity avoided.

To do this, imagine the balls numbered consecutively from 1 to 100 and consider two gambles of the type considered above but contingent on the numbering of the balls and not their colour. In the first you get the prize only if ball 37 is drawn: in the second, if ball 81 is drawn. Then the requirement is satisfied if you are indifferent between these two gambles, whatever be the different numbers chosen; for then you think ball 37 is just as likely as ball 81, and then for any ball. In other words, the same concept of indifference between gambles is used to ensure each ball has the same chance of being drawn as is then used to define probability. The basic concept is indifference. The ball is said to be drawn *at random*.

Objections have been raised because the standard involves gambling and some people object to gambling. The confusion here is due to inadequacies in the English language (or in my use of it). We are all faced with uncertain events like 'rain tomorrow' and have to act in the reality of that uncertainty—shall we arrange for a picnic? We do not ordinarily refer to these as gambles but what word can we use? In this sense all of us 'gamble' every day of our lives, and the word is being used in this sense. The gambles that people object to are unnecessary gambles on horses, or sport, or cards, usually conducted for monetary gain or excitement. The prize in our case need not be awarded: it is only contemplated. The essential concept is *action* in the face of uncertainty. How would you act in choosing between the events B and E? The number of balls was 100 for simplicity, giving probability to two decimal places or as an integer percentage. Any number, N, could be used when the probability would be b/N for the selected b to achieve indifference.

Notice that the example used an event E of rain tomorrow having the property that in time, after tomorrow, it will be known whether the event is true or false and the prize can be awarded, or not, settling the gamble. In the case of some events such a settlement cannot be guaranteed: for example, that Shakespeare wrote the plays. Here the gamble on E is unlikely ever to be settled. The difficulty is not too serious since your attitude to the gambles need only to be contemplative rather than involved. Also the same objection could be raised about light as a standard for distance: it is not directly available for two objects not in line of sight. There it can be avoided by measuring other distances and using triangulation. After the rules of probability have been derived it will be seen (section 9.8) that a similar device may be used for uncertainty. The Shakesperian example will be considered again in section 9.7.

In section 2.3 a distinction was made between statistical and non-statistical events; essentially dependent on the property of repetition. Notice that the standard uses no repetition and is non-statistical. The ball is only drawn once from the urn to settle the gambles, after which the whole apparatus could be destroyed. To emphasize this point suppose drawings were repeatedly made, the selected ball being replaced in the urn before another was withdrawn, and that a count was made of the proportion of black balls actually seen; would it be $b/100$, the proportion in the urn? Not necessarily, for it might be that black balls are more readily selected than white ones (the black paint on the naturally white-metal balls makes them easier to grip). This would not affect the standard nor the value of the probability which only mentioned what you thought, not what was actually true.

2.7 THE NATURE OF PROBABILITY

This last comment additionally makes the important point that the probability derived by reference to a standard depends on you, the person making judgements about all the balls being equally likely to be withdrawn, and about the relative merits of the two gambles. We say that the probability is *subjective*: it depends on the subject making the judgements. There is no reason to think that two people contemplating rain in London tomorrow would come up with the same value of b. Americans seem to think that it is always raining in London and would doubtless have a larger value than a Highland Scot, who considers London dry relative to his homeland. To put the point differently: probability does not have an existence outside of a person, as does the length of a table. Probability expresses a relationship between a person and the world he is contemplating. Some have regarded this as a disadvantage. In fact, the subjective nature of probability is its great strength, for it describes a real situation of a subject, or observer, contemplating a world, rather than talking about a world divorced from the observers living in it, as do some other sciences.

Recognizing that two observers may have different probabilities for the same event, the suggestion has been made that the difference arises not because the people are different but because they have different amounts of information, like the American and the Highland Scot in our example, and that if they shared the same information they would agree; for example, if they both had meteorological data over the last 30 years giving the frequency of rainy days in London. But suppose the Scot knew it had rained yesterday and the American did not: and it was not any day but 17 May. It is very difficult, if not impossible, to make the informations available to two people exactly the same. Also, despite several attempts, no-one has been able to say what this impersonal probability should be for an event E with prescribed information.

The climatological data just mentioned would provide the chance (section 2.4) of rain tomorrow, not your probability. This chance is 42% but would certainly not be your probability if you knew it had rained yesterday. Incidentally, the value is much lower than most people, especially Americans, think. For comparison, in Washington it is 34%.

For a given subject contemplating an event the probability will depend on the information he has. Indeed one of the most important tasks is for us to see how the probability changes with information.

It is here advantageous to introduce some notation. We have already denoted an uncertain event by E. Let the information available to you when considering E be denoted by H. (As a mnemonic, H for history.) Finally write $p(E \mid H)$ for the probability of E with information H: we usually say, the probability of E, given H. $p(\ )$ is the symbol for probability. The vertical line means 'given'. Symbols in the brackets to the left of that line refer to the uncertain event whose probability is stated: those to the right refer to what is known. The notation does not refer to the person assessing the probability, though, as has been argued, it may depend on him. Our discussion will almost always refer to one person who need not therefore be mentioned. A brief discussion on many-persons situations is given in Chapter 10. A vital point to notice is that a probability depends on *two* things, the event E and the information H. It is nonsense to talk about the probability of E, though in the definition above we did just this to prevent the argument becoming too involved. We should have said 'the probability of rain tomorrow in London, given your present information'.

To summarize the argument so far: for a person with information H contemplating an uncertain event E, the uncertainty has been measured by a number called a probability and written $p(E \mid H)$.

2.8 COHERENCE

An important assumption has tacitly been made in the definition of probability through a standard. It has been assumed that any event E can be compared with an event B concerning balls in an urn. The hidden assumption is that this comparison can be made in a coherent manner. By this is meant that if E_1, E_2, and E_3 are three events where you think E_1 is more likely than E_2 and E_2 than E_3, then you necessarily think E_1 is more likely than E_3. The last comparison is said to *cohere* with the other two comparisons. That this needs to be true for our method to make sense is easily seen by considering the events B_1, B_2, and B_3 judged equivalent to E_1, E_2, and E_3. Since E_1 is more likely than E_2, b_1, the number of black balls in B_1, will exceed b_2. Since E_2 is more likely than E_3, b_2 will exceed b_3. Hence b_1 exceeds b_3, which will only be true if E_1 is more likely than E_3. Hence the coherence.

The assumption of comparability is justified by the point already made that no-one has been able to describe which events are comparable and which not; and by the consideration that complete comparison is desirable. The division into statistical and non-statistical fails. The assumption of coherent comparability is easily justified. Suppose a person did violate it by saying:

(1) E_1 is more likely than E_2,
(2) E_2 is more likely than E_3, yet said
(3) E_3 is more likely than E_1.

Suppose he is to receive a prize if E_3 occurs, otherwise he gets nothing. Then, keeping the prize fixed throughout the discussion, he would prefer, by (2), to base the receipt of the prize on E_2. So he would pay a sum of money to have

E_2 substituted for E_3. The same argument using (1) shows he would pay a further sum to have E_1 subsitituted for E_2. Finally, (3) shows that a third sum would be forthcoming to replace E_1 by E_3. But now we are back to the original gamble and he has parted with three sums. The cycle may be repeated if he holds to the statements (1)–(3) and the incoherent person is a perpetual money-making machine. Clearly, this is absurd and only coherent comparisons make sense.

The point may be thought to be rather trivial and obvious. Our excuse for raising it is that coherence, in a more general form, is the bedrock upon which our whole analysis of decision-making is based. We do not wish to say that any of the statements (1)–(3) above are right or wrong, merely that as a set they are wrong (or incoherent). Later it will not be possible to say that a decision is right but only that these decisions cohere, or not. It is the relationships between events or decisions that matter, not the individual events or decisions. The principal aim of this book is to show that these relationships are rather precisely defined and are certainly not arbitrary. Our approach is both very liberal and yet very restrictive. Liberal in that it allows for a wide range of preferences: restrictive in that these preferences must obey certain rules (of coherence). Much decision-making is wrong because the rules are violated.

The next chapter is concerned with the rules of probability, and the reader who is convinced with the arguments so far presented in this chapter may omit the rest of it and proceed directly there. In the remainder an alternative approach is given, which helps to reinforce the coherent viewpoint, and insurance is discussed some more. When so many results depend on basic ideas it is well to look at the ideas from several sides. When the different approaches yield the same results, the validity of those results is enormously strengthened. The new one will also expose a basic principle of decision-making.

2.9 HOW GOOD IS A PROBABILITY ASSESSMENT?

Suppose that for an event E on information H you had a probability $p(E \mid H)$, or simply p. Thus, if E is the event that a coin falls head, you might have $p = 0.5$: for the event that France is larger in area than Spain, you might have $p = 0.3$. The former event is, for you, more likely. A practical question we might ask is whether you had done the measurement correctly or whether you had got the number 'right'. This is certainly a question that might be asked, and answered, of the measurement of length. How might we answer it for an event? At first sight this seems difficult, but an ingenious way, available for at least some events, is the following. An event is either true or false and it would be natural to say the measurement was good if you had thought an event, subsequently shown to be true, had been likely, that is, given a large p; and an event later seen to be false to have been given a small p. Let us see how well this works by trying it on some events. In Table 2.1 are listed ten events. The reader is invited to associate with each event a number describing in some sense how likely it is to be true in his judgement. Each event is given

Table 2.1. Some uncertain events

1. The composer Michael Haydn was the father (brother) of Joseph Haydn, the more famous composer.
2. Claret is the English name for the wine of Bordeaux (Burgundy).
3. Charlotte (Emily) Bronte wrote *Jane Eyre*.
4. The oldest national anthem is that of Great Britain (France).
5. The modern viola has 5 (4) strings.
6. Zambia is the modern name of former Northern (Southern) Rhodesia.
7. Rome, Italy is south (north) of Washington DC, USA.
8. Perigee is the point farthest from (nearest to) the earth in the orbit of an earth satellite.
9. In the 1970s the world production of wheat was about 5 (9) billion bushels.
10. Goodyear (Firestone) was the inventor of the vulcanization process that made possible the commerical use of rubber.

in an alternative form in parentheses; one of the two forms is assuredly correct,* you are asked to give your probability for the unbracketed form. Remember, the number is to be bigger the more likely the event is to be true. To achieve standardization two extra conditions are imposed. Some events in Table 2.1 you may know to be true; they are the most likely of all events. Let them be given the value 1. Some events are judged false: assign the value 0 to them. In other words, associate with each of the events a number between 0 and 1. At this stage you don't know how to do this, all you have is a feel for their truth. Just do the best you can.

Having done this, turn to the answers to the exercises at the end of the book. Amongst those will be found a list saying which events in Table 2.1 are true, which are false. Now compare your numbers, the values p, with the truth statements. You will feel that a good job has been done if the true events have large numbers and the false ones small numbers. For the event above, that France is larger in area than Spain, the value of $p = 0.3$ is bad since France does have the greater size. At the extremes you will feel a major mistake has been made if $p = 1$ for a false event; that is, one you thought was true but indeed was false; and equally with $p = 0$ for a true event. Less extreme, $p = 0.8$ for a true event will be better than $p = 0.6$. How much better?

2.10 SCORING RULES

A way of answering this question is by means of a *scoring rule*. This is a rule which gives each measurement p a score depending on whether the event is true or false, the score measuring the quality of the measurement p. The particular score proposed is $(1-p)^2$ if the event is true and p^2 if it is false. The score is to be thought of as a penalty score, so that the smaller the score the better

* The authority is the current (1984) edition of *Encyclopaedia Britannica*.

you have done. It is usual to multiply the values by 100 and to ignore values after the decimal point, if any. It is called a quadratic scoring rule. To illustrate, consider an event subsequently found to be true. The value $p = 1$ means that you were correct in thinking it was true and receive no penalty. The value $p = 0.9$, meaning you thought it highly likely, gives a modest score of $(1 - 0.9)^2 = (0.1)^2 = 0.01$ which, on multiplying by 100, gives only 1. A lower value, $p = 0.7$, gives a penalty of 9 and the opinion that it was just as likely to be true or false, $p = 0.5$, gives a larger 25. A low value like $p = 0.2$ produces 64 and thinking it false, $p = 0$, yields a maximum score of 100. Table 2.2 gives a list of the penalties associated with various values of p for true and false events.

Now turn to the p-values you gave in response to the events in Table 2.1, score them according to the quadratic rule of Table 2.2, and add the ten values. How well did you do? The answer to this question depends on two things. First, it depends on the property being investigated, your ability to express your uncertainty numerically: second, it depends on how much you know. Someone with a great fund of knowledge and a memory to retain it will do better than a reader with little understanding of the questions. Some idea of reasonable scores can be obtained by considering extreme cases. If all the answers were known to you and all were correct the total score of 0 results: whereas if all were wrong there is a penalty of 1000. These are the limits. Suppose you felt indifferent about each of the events, or even thought the whole thing ridiculous: there are two things you might do; give each $p = 0.5$, or guess putting $p = 1$ for those you guess to be true and $p = 0$ for the others. The former procedure will give 25 for each event, a total of 250. The latter will give 0 or 100 for each event and if half are right—a reasonable assumption if guessing—a total of 500. We immediately see that the assignment of $p = 0.5$ is a much better strategy, giving half the score with guessing. Hence in your

Table 2.2. Quadratic scoring rule

Probability	Score: True event	Score: False event
0.0	100	0
0.1	81	1
0.2	64	4
0.3	49	9
0.4	36	16
0.5	25	25
0.6	16	36
0.7	9	49
0.8	4	64
0.9	1	81
1.0	0	100

The score is $(1 - p)^2$ for a true event, p^2 for a false one.

response to the questions of Table 2.1 you ought not to have scored more than 250, since you do have some knowledge of the events. By analysing the ten individual scores you can see what the quadratic rule is doing. If an event is true, a value of p in excess of 0.5 will give a modest score, it is the small values of p that make a substantial contribution. Thus $p = 0.7$ gives 9 but $p = 0.3$ (as with the question about France and Spain) gives 49. The reason for not selecting extreme values of p, near 0 or 1, is that they will result in extreme penalties if the truth does not lie in the direction you thought. Someone who is fairly confident the event is true, but is wrong, will get a huge 100 if he gives $p = 1$ but only 81, a reduction of 19 if he reduces to $p = 0.9$. Whereas if he is correct the reduction in score as p rises from 0.9 to 1 is only 1: it pays to be cautious. On the other hand, too much caution does not pay. Consider someone who thinks the event is true but is not at all sure and assigns $p = 0.6$. If correct he will score 16 but this can be almost halved by going to $p = 0.7$ and getting 9. A knowledgeable person, lacking in confidence, will give values around $p = 0.5$ but moving away from $\frac{1}{2}$ in the right direction. This score could be reduced by greater departures from $\frac{1}{2}$. The confident person will go for the extreme p's. He will do well if his confidence is justified but will come a cropper if not.

2.11 SCORING RULES AS AN AID IN ASSESSMENT

Now that you have seen something of the consequences of the scoring rule try the 10 similar questions in Table 2.3, bearing in mind the rule, check with the answers at the end of the book and then score them. Hopefully, your score will have been reduced because of your awareness of the quadratic rule. You will have become better at the measurement of uncertainty. Of course, this need not happen. The second set of events in Table 2.3 may be harder for you than those in Table 2.1 or just chance, with as few as 10 questions, may have acted against you. But generally experience shows that the rule is useful in improving people's appreciation of uncertainty.

Table 2.3. More uncertain events

11. Johann Strauss the Younger (Elder) wrote the *Blue Danube* waltz.
12. Hock is the English name for some wines from Alsace (Germany).
13. *The Golden Bough* was written by James Frazer (Henry James).
14. The potato was introduced into Europe from China (America).
15. The modern oboe has a single (double) reed.
16. Guyana is the modern name of British (Dutch) Guiana.
17. Santiago, Chile is west (east) of New York City, USA.
18. The mean surface temperature of Venus is about 450°C (250°C).
19. About 22 (15)% of the world's population is Muslim.
20. The first process for making steel inexpensively was first invented by Bessemer (Kelly).

The quadratic rule has been used in the United States in the training of weather forecasters. The event considered is 'rain tomorrow' at a prescribed place and the forecaster is required to give his p-value or probability. This is repeated for a month or more and the total quadratic score is recorded. The better the meteorologist, the lower his score. The training is reflected in the forecast provided on American television, which will often take the form 'there is 85% probability of rain tomorrow', meaning $p = 0.85$. It is clear that the training is not always as good as it might be because forecasters are often confused over what probability means. One was heard to say that it meant there would be rain over 85% of the area served by the television station. Unfortunately, probability ideas have not been used in Britain where the forecaster says it will rain or uses other, vague, literary expressions like 'there is a possibility of rain'.

The habit of making positive statements like 'it will rain' when the true situation is one of uncertainty is deplorable. I have just read an article on nuclear strike which says that if NATO does this 'Russia will . . .'. How does the author know what the Soviet Union will do? Later on he uses the phrase 'all the probability is that': presumably this means there is a little doubt in his mind, but not much. Politicians are not the only offenders, the concept of certainty pervades religion when really all is doubt. The habit of replacing uncertainty by definitive assertions arises because of our deep dislike of uncertainty and the desire to feel we are in control and know what is happening. It is reinforced by the instruction received at school. Faced with a series of questions, Johnnie is expected to know the answer and all he is allowed to say is 'true' or 'false'. We saw above how guessing true or false would give a score around 500 whereas honest ignorance and $p = 0.5$ produces about half this value. In examination by multiple-choice questions the candidate is given five answers and asked to mark the correct one. What he ought to do is to give his probabilities for each of the five suggestions.

2.12 SCORING RULES AND REPEATED EVENTS

It was explained above that some events naturally have their uncertainties measured, particularly those occurring in games of chance. There is a connection between such statistical events and the scoring rule approach that is now explained. The explanation will reinforce the point just made that guessing is bad. Suppose you had a die and the event considered is that a 5 or a 6 will show when the die is rolled. If the die is fair the chance, or long-run frequency, is 1/3, and it seems natural to say $p = 1/3$ for the probability that the event will occur. Let us see what would happen if the die were rolled 27 times and each time you said $p = 1/3$. On about 1/3 of occasions, 9 times, the event would occur and you would score $(2/3)^2$, or 44: on the remaining 18 the score would be $(1/3)^2$, or 11. The total would be $396 + 198 = 594$, or an average of 22 per event. This is the lowest score it is possible to obtain. Consider, for example, a guessing strategy, sometimes saying $p = 1$, sometimes $p = 0$. This strategy is effectively forced on you when you have to decide whether to bet ($p = 1$) or not ($p = 0$). The score will either be 100 (when wrong) or 0 (when correct). Suppose, as seems reasonable, you guess the 5 or 6 will appear ($p = 1$) 1/3 of the time. Consider again what happens over 27 rolls of the die. In 1/3, 9, of these the event will occur, on 1/3 of those, 3, you will be right and get a zero score: in the 18 others the event will not occur and on 2/3 of those, 12, you will get zero. Thus $3 + 12 = 15$ out of the 27 will produce zero. But on the remaining 12 you will be wrong and the score will be 100. The total for the 27 rolls will be 1200, an average of a little over 44 per event. This is twice what the use of $p = 1/3$ consistently will give. Guessing is a very bad strategy. Readers with the necessary mathematics can prove that no strategy improves on $p = 1/3$. Other readers may like to do the arithmetic for any strategy that occurs to them. For example, I have heard it argued that since we do not know whether a 5 or 6 will occur or not we should say $p = 0.5$. The average score per event is then 25, a little higher than the 22 with $p = 1/3$. This example demonstrates that the scoring-rule device for measurement gives a result that agrees with that based on chance whenever the statistical concept is appropriate.

We now have two approaches to the numerical description of uncertainty: one based on direct comparison with a standard, the other using a scoring rule to judge the quality of the assessment, however it was done. We next look briefly at a third method, based on insurance. This has two advantages: it relates closely to the decision process which is the main study of this book, and it uses already existent expertise.

2.13 INSURANCE

Consider some uncertain event, E, which may or may not happen to you. We shall think of E as a disastrous event, that is, one which you strongly hope will not happen; but the argument works equally well if E is favourable, the signs of some of the quantities being reversed in that case. For example, E may concern the loss of some valuable item, like a camera. Suppose it would cost you an amount c (in suitable units, be they pounds, francs or dollars*) to recover from E were it to happen. Thus it might cost c dollars to replace the lost camera. If your capital is C dollars there are two possibilities: either E does not occur and your capital remains at C, or E does occur and your capital drops to $C - c$. In either event you will be in the same position in all other respects except the possible loss of capital: in particular, you will have a camera. Faced with a possible loss of c dollars, you may contemplate taking

* There is no agreed international unit for money. Throughout this book we have called the unit a 'dollar' without any suggestion that it is a US or any other nation's unit. The reader is free to interpret it as he wishes. He will probably not be able to get away with a single interpretation throughout the book: to make some examples realistic to him he will have to enlarge or diminish its real value in his mind.

out an insurance policy which will pay you c dollars in the event of E occurring. If the premium on the policy is m, acceptance of the insurance means that your capital is certain to be $C-m$ whether you lose the camera or not, and the uncertainty of E has been removed. The situation is conveniently described in tabular form (Table 2.4). $\bar{E}$ means the opposite event to E.

Typically, m will be much less than c. Now, for sufficiently small premiums the insurance would be attractive whilst for large premiums it would be better to bear the risk oneself. By an argument similar to that used in section 2.5, there must be a unique value of m such that premiums smaller than m would be acceptable, and those larger than m would not. In other words, for that value, m, of the premium, insurance and non-insurance are equally attractive. Notice, by the way, that such a comparison of insurance with non-insurance is forced on you by the need to make a decision about the policy. A comparison must be made: there is no way out. This pressure was not present when comparing uncertain events in section 2.5.

Now consider another situation. Suppose your capital is $C-m$, the value m being that unique borderline value found in the preceding paragraph, and someone offers you the following gamble for a stake of $c-m$: if a *white* ball is drawn from the urn (the event $\bar{B}$) you will win an amount m (and your stake will be returned), otherwise you will lose your stake. Thus if B does not occur your capital will rise to C: if it does it will drop to $C-c$. Table 2.5 describes this situation. For a small value of b, the number of black balls in the urn, you would accept the gamble, for a large value you would refuse it. Again, there will be a value b and a unique number $p(B)$, the probability of B, which is critical in the sense that values smaller than b would make you accept, larger than b would make you reject. In other words, for that value b, the acceptance and the refusal of the gamble are equally attractive. Again notice that in view of the decision forced upon you, such comparisons must be made and expressed by your action in accepting or refusing the gamble. If we compare the tables for insurance and for the gamble we see that the situations if the gamble is refused or the insurance taken are identical, namely a certain capital of $C-m$. These are respectively just as attractive as the situations if the gamble is accepted or the risk if the insurance is refused, since in both cases the choices are finely balanced. Consequently the gamble and the risk of not insuring are themselves equally attractive. Whether the gamble is accepted or the risk taken, the outcomes are either $C-c$ or C in both cases, the only difference being that in one case the event E and, in the other, the event B determine which it is to be. Consequently there is no escaping the con-

Table 2.4. Insurance against E

Uncertain event	E	$\bar{E}$
Outcome without insurance	$C-c$	C
Outcome with insurance	$C-m$	$C-m$

Table 2.5. Gamble on R

Uncertain event	B	$\bar{B}$
Outcome if gamble accepted	$C-c$	C
Outcome if gamble refused	$C-m$	$C-m$

clusion that E and B are equally uncertain, and that therefore the probability of E must be equal to $b/100$.

This procedure, involving considerations of capital and costs, is complicated and would only be recommended to a decision-maker experienced in insurance. In the next chapter we shall use the concepts of probability derived from a standard and from the use of a scoring rule and show that both lead to the same laws of probability that contain the essence of the idea of coherence.

Chapter 3

The Laws of Probability

'"It's all chance, but we can't stop now"'

Maid in Waiting, Ch. 28.

3.1 THE CONVEXITY LAW

In the last chapter it was shown, using two different methods, how a numerical description of uncertainty could be obtained. The methods used respectively a standard and a scoring rule. Our task in this chapter is to show that the numbers, the probabilities, however produced, obey certain laws. The analogy with length is again helpful. In determinations of size and distances we use measuring devices like micrometers, theodolites, telescopes: all of which are based on principles derived from the laws of geometry. The measuring devices for uncertainty are similarly based on the laws of probability. Without probability, life as we know it would be impossible. Chance is the basis of the laws of genetics and if ever laws of sociology are obtained, they will be found to be probabilistic. These laws* are therefore of basic importance yet simply follow from the ideas already discussed.

We saw in Chapter 2 that a person contemplating an event E with information H has probability $p(E \mid H)$. Our first result is the convexity law.

Convexity law

$p(E \mid H)$ lies between 0 and 1

In mathematical notation

$$0 \leq p(E \mid H) \leq 1$$

This law is immediate from the assessment of probability using a standard because the value was the proportion of black balls in the urn, necessarily between 0 and 1. The law is to some extent a convention. We could, for example, use 100 times the proportion, as when a percentage is used. If so, the

* In the first edition of this book the laws were numbered. Here they are given names as an aid to the memory. The convexity law was not numbered previously.

multiplication law (section 3.5) would have a factor of 100 in it but otherwise be unaltered. Some have argued that probabilities with arbitrarily large values should be used, but this causes problems that will not be considered in this book. The extreme values, 0 and 1, have special significance that will appear when Bayes' theorem is discussed in section 6.7.

The convexity law is not quite obvious when the quadratic scoring rule is used and it is worth studying the demonstration, because to do so exposes another aspect of coherence. We show that the choice of any number not between 0 and 1 would be absurd. For simplicity in exposition, write p for $p(E \mid H)$. Suppose, for example, $p = 2$ is selected. If E is true, the penalty is $(2-1)^2$, or 100 on scaling. A false event will yield a score of $(2)^2$, or 400. Thus $p = 2$ can give 100 or 400. Contrast this with what happens when $p = 1$ is stated: the possible scores are 0 or 100. In both cases the score is reduced: if E is true, from 100 to 0; if false, from 400 to 100. Consequently $p = 1$ is inevitably better than $p = 2$. This argument will work for any value of p in excess of 1. A similar result applies for p negative, $p = 0$ is always better. It does not work if p lies between 0 and 1: then a change in p will increase one penalty and decrease the other. Thus $0 \leq p \leq 1$ and the law is proved.

Let us isolate the principle used here. Look upon the choice of p as a decision. Let d_2 be the decision to use $p = 2$, d_1 that to use $p = 1$. Then the outcome of d_2 when E is true is worse than that of d_1; and the same applies when E is false. Or the outcome of d_2 is worse than that of d_1 whether E is true or false, in other words irrespective of E, and therefore d_2 is worse than d_1 when E is uncertain. This is the *sure-thing principle*. If d_2 is worse than d_1, both when E is true and when E is false, it is worse when E is uncertain.

This is a second example of coherence between three judgements all concerning d_2 and d_1; one when E is true, one when false, and one when uncertain.

3.2 THE ADDITION LAW

We begin with a few preliminaries. Remember that two events, E_1 and E_2, are *exclusive* if they cannot both occur (section 1.6.). For *any* two events E_1 and E_2 consider the event that occurs if either E_1 or E_2 occurs, and only then. We write it 'E_1 or E_2' and call it the *union* of E_1 and E_2. Thus if E_1 is the event that a hand of two cards consists of an ace and a court card, and E_2 that it contains an ace and a ten; E_1 or E_2 is the event variously called blackjack or pontoon. Suppose that you have assessed $p(E_1 \mid H)$ and $p(E_2 \mid H)$; that is, the probabilities of two events on the *same* information; and then go on to consider the probability of the union, $p(E_1 \text{ or } E_2 \mid H)$. The addition law says that if E_1 and E_2 are exclusive, the latter is the sum of the two probabilities for the separate events.

Addition Law

If E_1 and E_2 are two *exclusive* events, then on information H

$$p(E_1 \text{ or } E_2 \mid H) = p(E_1 \mid H) + p(E_2 \mid H) \tag{3.1}$$

It is convenient to introduce a convention which can be misleading but is often valuable in simplifying expressions. In the addition law H, the information, is the same in all probabilities which occur. It is therefore reasonable to omit explicit reference to H, understanding that every event is considered conditional on H. Then the law says

$$p(E_1 \text{ or } E_2) = p(E_1) + p(E_2) \tag{3.2}$$

We first provide some examples of the use of this law. Suppose a die, which is judged to be fair, is rolled in the usual way—these conditions form part of H. Let E_1 be the uncertain event that the die will come to rest with a 1 uppermost. Let E_2 similarly refer to a 2 being uppermost. Then typically $p(E_1) = p(E_2) = 1/6$. By the theorem, the probability of getting a 1 or 2 uppermost is therefore $1/6 + 1/6 = 1/3$. E_1 and E_2 are certainly exclusive since you cannot get both a 1 and a 2 showing. The theorem would still apply even if the die was not thought to be fair. For example, suppose you thought the higher numbers were more likely, you might choose $p(E_1) = 95/600$ and $p(E_2) = 97/600$ (both a little less than 1/6). We could still add the probabilities and obtain the probability of a 1 or a 2 as $(95 + 97)/600 = 192/600$. In other words, the result holds for any H and for any coherent assignment of probabilities to the 6 exclusive and exhaustive numbers that might be uppermost.

Consider again our friend contemplating crossing the mountains in winter (section 1.9). Let E_1 be the event that the pass is open and he gets through without an accident. Let E_2 be the event that the pass is open and he has an accident in using it. Suppose he assesses $p(E_1) = 3/4$ and $p(E_2) = 1/20$. These events are certainly exclusive, he cannot both have and not have an accident, and the event E_1 or E_2 is simply the event that the pass is open (since he either will, or will not, have an accident). Consequently it follows that he must assess the probability of the pass being open as $3/4 + 1/20 = 16/20 = 4/5$.

The danger with the omission of H is that its presence may be forgotten. We now proceed to prove the addition law on the basis of the measurement of probability by reference to a standard.

3.3 PROOF OF THE ADDITION LAW

The first probability $p(E_1)$ will be assessed by choosing b_1 balls in the urn of N balls to be black and having $p(E_1) = b_1/N$. Similarly, $p(E_2)$ will correspond to b_2 balls being blue say, a distinguishing colour, and $p(E_2) = b_2/N$. So the extraction of a black ball corresponds to E_1 and of a blue one to E_2. But since E_1 and E_2 are exclusive, no ball can be painted both black and blue since its extraction would correspond to both E_1 and E_2 occurring, which is impossible. Hence there are b_1 black, b_2 blue, and $N - b_1 - b_2$ white balls in the urn and the event 'E_1 or E_2' corresponds to a coloured ball being drawn of which there are $b_1 + b_2$. Consequently $p(E_1 \text{ or } E_2) = (b_1 + b_2)/N$, as the law says.

An important special case of the law arises when E_2 is the negation of E_1: the event which occurs if and only if E_1 does not. The negation of the event of a die showing an even number is that it shows an odd one. We write $\bar{E}$ for the negation of E. Since E and $\bar{E}$ are exclusive and their union is certain to occur, and therefore has probability one,

$$p(E) + p(\bar{E}) = 1$$

or

$$p(\bar{E}) = 1 - p(E) \tag{3.3}$$

In words, the probability that an event does not happen is one minus the probability that it does.

It is important to remember the requirement in the addition law that E_1 and E_2 be exclusive. The proof demonstrates the need for the condition in removing the possibility that some balls be both blue and black and hence that the number of coloured balls be less than $b_1 + b_2$. This remark shows that, in general, $p(E_1 \text{ or } E_2)$ is less than $p(E_1) + p(E_2)$: just how much less will be seen when another event combination has been discussed.

The addition law also applies to any number, n, of exclusive events and says that the probability that one of the events occurs is the sum of the n individual probabilities. In symbols

$$p(E_1 \text{ or } E_2 \text{ or } \ldots E_n) = p(E_1) + p(E_2) + \ldots + p(E_n)$$

Here the three dots as before mean 'and so on up to'. The proof is again easy, but it uses a technical mathematical trick. The trick is worth looking at since we often want to use it. The result

$$p(E_1 \text{ or } E_2) = p(E_1) + p(E_2)$$

holds for *any* two exclusive events, so let us use it with E_2 replaced by a different event, namely E_2 or E_3. This is still exclusive of E_1. It will then read

$$p(E_1 \text{ or } E_2 \text{ or } E_3) = p(E_1) + p(E_2 \text{ or } E_3)$$

where every E_2 in the first result has been replaced by E_2 or E_3 in the second. But

$$p(E_2 \text{ or } E_3) = p(E_2) + p(E_3)$$

by a second application of the result, this time with E_1 replaced by E_3. This is legitimate since E_2 and E_3 are exclusive. Combining the last two equations we have

$$p(E_1 \text{ or } E_2 \text{ or } E_3) = p(E_1) + p(E_2) + p(E_3)$$

This principle can easily be extended to any value of n; that is, to any number of events. The mathematical trick is to use a single result repeatedly for different events, and then to combine the different expressions so obtained.

An immediate consequence of this is that in our list of exclusive and exhaustive events introduced in Chapter 1, $\theta_1, \theta_2, \ldots \theta_n$ the probabilities $p(\theta_1), p(\theta_2), \ldots p(\theta_n)$ must add to 1 since, by exhaustion, their union is certain. This might not always happen on a first attempt at assessment and the addition law provides another example of coherence in requiring that they do: that the n

numbers cohere in the sense of adding to one. For example, suppose you are contemplating how you might die. Let θ_1 be heart attack, θ_2 cancer, θ_3 violence, θ_4 some other non-violent cause. Then since these events are all distasteful to you the probabilities may well add to less than one. If this happens, coherence, through the addition law, requires that at least one of them be changed.

The next section contains a proof of the addition law on the basis of a quadratic scoring rule and illustrates the force of the sure-thing principle. It may be omitted without loss of continuity.

3.4 SCORING RULE PROOF OF THE ADDITION LAW

We prove the special case $p(E) + p(\bar{E}) = 1$ involving an event E and its negation $\bar{E}$. Suppose you gave the value x when asked to consider E, and y for $\bar{E}$. Then your total score if E was subsequently found to be true would be the sum of $(x-1)^2$, since E is true, and y^2, since $\bar{E}$ is false: that is, $(x-1)^2 + y^2$. On the other hand, if E is false, $\bar{E}$ is true and the total score is $x^2 + (y-1)^2$. What we want to show is that if these two values, x and y, do not add up to 1 then both these scores can be reduced by changing x and y so that they do add to 1. If x and y do not add to 1, take the deficiency $1-(x+y)$, half it, call it t, and add to both x and y, giving $x+t$ and $y+t$, now adding to 1. So $t = (1-x-y)/2$ and may be negative. Consider the score when E is true, $(x-1)^2 + y^2$. With the new values it becomes $(x+t-1)^2 + (y+t)^2$. We claim this value is necessarily less than the original one. The difference is

$$\begin{aligned} &(x-1)^2 + y^2 - (x-1+t)^2 - (y+t)^2 \\ &= -2t(x-1) - 2ty - 2t^2 \\ &= -2t(x+y-1+t) \\ &= 2t^2, \text{ since } x+y-1 = -2t \end{aligned}$$

This is positive, so that the change from (x, y) to $(x+t, y+t)$ does reduce the score. Similar calculations for the score when E is false establish that it too is reduced. Hence by the sure-thing principle $x+y$ must add to 1.

A numerical example may prove instructive. Suppose $x = 0.5$, $y = 0.4$; the deficiency is 0.1 and t is 0.05, so that the suggested new values are 0.55 and 0.45. For E true the original score was $(0.5)^2 + (0.4)^2 = 0.41$: with the new values it is $(0.45)^2 + (0.45)^2 = 0.405$, a reduction. For E false the change is from $(0.5)^2 + (0.6)^2 = 0.61$, to $(0.55)^2 + (0.55)^2 = 0.605$, another reduction.

The proof that the general form of the addition law follows from the special case of an event and its negation has to wait on some further results: see section 3.13.

3.5 THE MULTIPLICATION LAW

We have seen that two events, E_1 and E_2, can combine to form the union 'E_1

or E_2'. Another way they can combine is to form the intersection 'E_1 and E_2'. This is the event which occurs if and only if both E_1 and E_2 occur. If E_1 is the event of throwing an even number with a die, and E_2 is the event of throwing a multiple of 3, then 'E_1 and E_2' is the event of throwing a 6. If E_1 and E_2 are exclusive, the case considered in the addition law, the event E_1 and E_2 is impossible.

The addition law only applies when dealing with exclusive events. The multiplication law holds for any two events and is trivial when the events are exclusive. The addition law involved the union 'E_1 or E_2': the multiplication law involves the intersection 'E_1 and E_2'. If E_1 and E_2 are not exclusive and can occur together, we might be informed about their occurrence in order, say E_1 first and then E_2. To illustrate, consider again the retailer wondering how many items to order (section 1.6). He may judge that his sales will depend on whether the government relaxes the conditions on borrowing or not. The relaxation is an uncertain event which we denote E_1. The retailer is faced with E_1 and some event concerning his sales: let us take as an example, E_2, the event of selling fewer than four items, which two events are certainly not exclusive. It may happen that the government does relax the conditions—that is, E_1 occurs—and that the retailer is informed about this. He will then no longer be interested in $p(E_2)$, but in $p(E_2 \mid E_1)$, the probability of selling fewer than four items, given a relaxation in restrictions on borrowing money. It is this probability that is involved in the multiplication law.

The law says that for any two events E_1, E_2 the probability of them both occurring on information H is the product of the probability of E_1 on H and the probability of E_2 on the information E_1 and H.

Multiplication law

For any events E_1, E_2 and information H,

$$p(E_1 \text{ and } E_2 \mid H) = p(E_1 \mid H)p(E_2 \mid E_1 \text{ and } H) \tag{3.4}$$

If explicit reference to H is omitted,

$$p(E_1 \text{ and } E_2) = p(E_1)p(E_2 \mid E_1)$$

The most common use of this law lies in calculating the probability of two events both occurring. To do this we need to know the probability of one of them, and the probability of the other, given the first. If our friend decides to cross the mountain in his car, what is the probability of his getting across without an accident? For this to happen, two events must both take place: E_1, the event that the pass is open; and E_2, the event that he does not have an accident. Suppose he assesses the first at 4/5, as suggested with the addition law. In symbols, $p(E_1) = 4/5$. Now consider $p(E_2 \mid E_1)$, the probability that he does not have an accident given that the pass is open: that is, the probability that he gets safely across the open pass. It may be easier to think about the probability of having an accident: there is something more positive about this than about the safe passage. By the addition law this will be simply one minus the former probability:

$$p(\bar{E}_2 \mid E_1) = 1 - p(E_2 \mid E_1)$$

(This follows from equation (3.3) above: E of that equation being replaced by E_2, and both probabilities being conditional on E_1.) Consideration of this value will depend very much on what is meant by an accident: for example, whether a skid that ends up by one facing in the wrong direction, but on the road and otherwise unharmed, is an accident. Suppose that on some reasonable definitions the value of $p(\bar{E}_2 \mid E_1) = 1/16$. This is to be interpreted as meaning there is just as much chance of an accident as getting four heads on four tosses of a fair coin. This is rather a high value, but we must bear in mind the hazardous conditions involved in crossing the high mountains. It then follows that $p(E_2 \mid E_1) = 1 - 1/16 = 15/16$, the chance of not having an accident. Inserting this value, together with that for $p(E_1) = 4/5$, into equation (3.4) of the multiplication law we immediately obtain (omitting H)

$$p(E_1 \text{ and } E_2) = 4/5 \times 15/16 = 3/4$$

Hence if the chance of the pass being open is 4/5 and the chance of getting safely across the open pass is 15/16, the chance of the pass being open and getting across is 3/4. We will return to this calculation again but first we prove the multiplication law using the standard of balls in an urn.

3.6 PROOF OF THE MULTIPLICATION LAW

As with the proof of the addition law, let E_1 correspond to b_1 black balls in the urn of N balls and E_2 to b_2 blue balls, but this time since E_1 and E_2 are not exclusive there is a possibility of balls coloured both black and blue corresponding to E_1 and E_2 being both true. How many will there be with both colours? Considering the union 'E_1 and E_2', the number will be b_{12}, where $b_{12}/N = p(E_1 \text{ and } E_2 \mid H)$. So the urn will have b_1 black balls of which b_{12} will also be blue. Now obviously

$$\left(\frac{b_{12}}{N}\right) = \left(\frac{b_1}{N}\right)\left(\frac{b_{12}}{b_1}\right)$$

$b_{12}/N = p(E_1 \text{ and } E_2 \mid H)$, $b_1/N = p(E_1 \mid H)$; what about the other factor (b_{12}/b_1)?

It is a basic property of the urn that you think each ball is as likely to be drawn as any other. This is true even amongst the b_1 black balls, of which b_{12} are also blue. Consequently if, in addition to H, you knew the extracted ball was black, the probability that it is blue as well is b_{12}/b_1, the proportion of blue balls amongst the black. This corresponds to knowing E_1 in addition to H and then assessing the probability of E_2. Hence $b_{12}/b_1 = p(E_2 \mid E_1 \text{ and } H)$ and the multiplication law is proved.

All the proof really uses is the fact that the proportion of balls that are both black and blue is the proportion that are black times the proportion of black balls that are blue.

Just as the addition law can be extended from two events to any number, so can the multiplication law. We content ourselves with stating it for three events, the extension to any number will be clear. Omitting H,

$$p(E_1 \text{ and } E_2 \text{ and } E_3) = p(E_1)p(E_2 \mid E_1)p(E_3 \mid E_1 \text{ and } E_2) \qquad (3.5)$$

The proof uses the same mathematical device as with the addition law in section 3.3, first using the law with two events E_3 and 'E_1 and E_2', and then with E_1 and E_2. Notice the last term on the right-hand side involves the information that E_1 and E_2 are both true (in addition to H). For n events the probability of E_n will be conditional on all the other events $E_1, \ldots E_{n-1}$ being true. This often makes the multiplication law difficult to apply but the situation is considerably alleviated by the condition of independence (section 3.14).

The multiplication law can also be proved on the basis of the quadratic scoring rule. That is, if you assess the three probabilities to be three numbers which do not obey the law then three other values can be found which do obey the law and for which the penalty score is reduced under all circumstances; that is, whatever be the truths if E_1 or E_2. The proof follows exactly the same lines as that of the addition law (section 3.4) constructing changes in the original values to lessen all the scores. The details are omitted.

3.7 COHERENCE

The convexity, addition, and multiplication laws are the three basic laws of probability. No one of them can be deduced from the other two, yet every other result in probability can be deduced from them without any further results. Probability is an enormous tripod built on these three legs. In some discussions they are treated as axioms: basic, intuitive, unproved results. In our presentation they have been themselves deduced from other, simpler concepts; either from a standard or by a scoring rule. (And there are other ways in which they can be deduced.) The key point here is that the ways in which statements of uncertainty can be combined are by no means arbitrary. One cannot sit down and think up apparently reasonable rules. The only ones are the three we have discussed: no more, no less. Some workers have suggested other types of combination: some lead to the concept of fuzzy sets. These ideas are false, because one is not free to engage in the intellectual exercise of law creation. The laws are forced upon you. It is a case of the inevitability of probability. The laws ensure that several statements of uncertainty cohere.

The laws constrain one's statements of uncertainty yet still leave a lot of freedom for personal views. It is like house-building: there are many designs of house yet they must all satisfy the constraints of the laws of geometry. Uncertainties are constrained by the laws of probability. These constraints typically provide the most useful way of assessing probabilities. Let us return to the mountain pass example to illustrate the point. Four events have been considered together with their probabilities:

Pass open, no accident	3/4	(= 0.75)
Pass open, accident	1/20	(= 0.05)
Pass open	4/5	(= 0.80)
No accident, given pass open	15/16	

(All the probabilities are, of course, conditional on the general circumstances surrounding the situation H, in addition to any others stated in the list.) The addition law says that the probabilities of the first two events must add up to give the probability of the third. The multiplication law requires that the product of the probabilities of the third and fourth events must equal the probability of the first. The two laws imply two relations between the four values. The reader may easily verify that, given any two of the four probabilities, the remaining two may be determined using the two laws. (There will have to be a few broad constraints on the given values: for example, if the first and third are given, the latter must not be smaller than the first, otherwise, by the addition law, the second probability will be negative, which is impossible by convexity.)

The lesson to be learnt from this is that when we come to the task of assigning probabilities we do not usually have to assign all of them, since many can be deduced from others using the probability laws. If this deduction leads to values that seem unacceptable then we must revise some at least of the original values to reach a set which both agree with our ideas and obey the laws. If anything has to give it is the initial assignments and not the laws, since the latter follow from our basic assumptions. For example, suppose our illustration had proceeded exactly as before up to the point where we considered the probability of an accident, given the pass was open. We used 1/16. Perhaps this was too high, and suppose we felt 1/24 was more reasonable. Then the fourth probability in the list would be $1 - 1/24 = 23/24$, and multiplying by the third as before we would have $23/24 \times 4/5 = 23/30$ for the first, and not 3/4. Consequently something would have to give and at least one of the probabilities be revised.

Another way of looking at the same situation is to think of the laws as assisting in the determination of the probabilities: they are parts of the apparatus for measuring uncertainty. We return to this topic in section 9.8.

In summary: the laws provide means whereby many probabilities can be calculated in terms of some basic values, thereby providing important checks on the original assignment of these basic values. No probabilities are any more basic than any others. As in our illustration, we can start from any convenient set. It is usually best to arrange the events in some convenient order and assign the probability of the first, then of the second given the first, and so on. In this way all probabilities can be deduced and the basic values can take any values between 0 and 1. Thus in our illustration the natural order is pass open, followed by accident: that is, the third and fourth probabilities in the list are assigned first and in that order. This avoids difficulties like the third in the list having to be greater than the first if they are regarded as the basic pair, but no general rule can be laid down and some situations have more convenient specifications.

We now prove two other 'laws'. They are called theorems because they can be deduced from the basic laws.

3.8 EXTENSION OF THE CONVERSATION

Theorem of the extension of the conversation

Let E_1 and E_2 be two events which are exclusive and exhaustive, and let A be any event. Then, omitting explicit reference to H,

$$p(A) = p(A \mid E_1)p(E_1) + p(A \mid E_2)p(E_2)$$

Since E_1 and E_2 are exclusive and exhaustive, E_2 must be the negation of E_1. So dropping the suffix from E_1 and writing $E_2 = \bar{E}$, we can alternatively state the theorem as

$$p(A) = p(A \mid E)p(E) + p(A \mid \bar{E})p(\bar{E}) \tag{3.6}$$

The original form permits extension to any number of events. Remember that all probabilities are conditional on the general circumstances, H, as well as any other conditions stated: thus, in full, $p(A \mid E)$ is properly $p(A \mid E \text{ and } H)$.

We illustrate its use by again using the example of the mountain pass. Let E denote the event of the pass being open, an event to which the probability 4/5 has already been assigned, that is $p(E) = 4/5$ and consequently $p(\bar{E})$, the probability of it being blocked, is 1/5. Let A denote the event of the driver having an accident. The value 1/16 has already been given to $p(A \mid E)$, the probability of an accident given that the pass is open. Consequently our previous considerations have associated probabilities with all the events on the right-hand side of equation (3.6) except $p(A \mid \bar{E})$, the probability of an accident given that the pass is blocked. Our laws do not enable us to deduce a value for this so we have to assign a number to it. If the pass is blocked presumably weather conditions upon the mountains are really bad and an accident, for example being stuck in a drift, is rather more likely than if the pass were free, so it might be reasonable to assign a higher probability to $p(A \mid \bar{E})$ than to $p(A \mid E) = 1/16$. Perhaps we might consider an accident twice as likely when the pass is blocked than when it is open: if so $p(A \mid \bar{E}) = 1/8$.

We now have all the numbers ready for the right-hand side of equation (3.6) and we can calculate $p(A)$, the probability of an accident, as

$$p(A) = 1/16 \times 4/5 + 1/8 \times 1/5 = 3/40$$

Hence if the decision is to drive, the chance of an accident is $7\frac{1}{2}\%$, say about the same as the chance that a random person was born in February.* Notice

* 3/40 = 27/360, not far off 28/365, or 29/366 if a leap year.

that this value, $3/40 = 6/80$, lies between the chance of an accident given that the pass is open, namely $1/16 = 5/80$, and the same chance given that it is blocked, namely $1/8 = 10/80$. A little thought will show you that $p(A)$ must always lie between $p(A \mid E)$ and $p(A \mid \bar{E})$. The proof is in the next paragraph and may be omitted without loss of continuity.

Suppose $p(A \mid E)$ is smaller than $p(A \mid \bar{E})$, as in our numerical example. We can certainly increase the right-hand side by replacing $p(A \mid E)$ by $p(A \mid \bar{E})$, but then the right-hand side equals $p(A \mid \bar{E})$ since $p(E) + p(\bar{E}) = 1$. Hence $p(A)$ is less than $p(A \mid \bar{E})$. Similarly we can decrease it by replacing $p(A \mid \bar{E})$ by $p(A \mid E)$ when it becomes $p(A \mid E)$. Hence $p(A)$ is greater than $p(A \mid E)$. Combining the two results, $p(A)$ lies between $p(A \mid E)$ and $p(A \mid \bar{E})$. A similar argument works if $p(A \mid \bar{E})$ is the smaller.

The extension theorem is perhaps the most widely used of all probability results and yet, at first glance, it looks as if it might rarely be useful, because it evaluates $p(A)$ by introducing another event E. What can be the point of adding E to the consideration of the chance of A occurring? The answer is that in many cases consideration of A alone is difficult and it is much easier to break it down into separate parts, given E, and given $\bar{E}$. Our mountain example is a case in point. The chance of an accident, A, depends on the weather conditions as does the state of the pass. Consequently it may be easier to assess the chances of an accident knowing the state of the pass than without this extra information. On the other hand, you might not find it so and may feel $p(A)$ can be directly assessed. That is fine, but if so, remember it must agree with equation (3.6). Thus if we already had $p(E) = 4/5$, $p(A \mid E) = 1/16$ an assignment of $1/20$ to $p(A)$ would mean that $p(A \mid \bar{E}) = 0$, since

$$1/20 = 1/16 \times 4/5 + p(A \mid \bar{E}) \times 1/5$$

only holds in those circumstances. Hence if you said $p(A) = 1/20$ you would have to live with $p(A \mid \bar{E}) = 0$; that is, there is no chance of an accident if the pass is blocked. You would probably think this unreasonable and be prepared to revise $p(A)$.

The fact that the extension theorem discusses an event A by introducing an extra event E (or, in the development which follows, several events) has given rise to the name of the theorem. One says that A is studied by extending the conversation to include E. We shall often have occasion to extend the conversation in this way.

As with the other two laws, the theorem can be extended to any number of uncertain events and not just two. In this form it reads that if $E_1, E_2, \ldots E_n$ are exclusive and exhaustive, and A any other event, then

$$p(A) = p(A \mid E_1)p(E_1) + p(A \mid E_2)p(E_2) + \ldots + p(A \mid E_n)p(E_n) \quad (3.7)$$

It is pretty clear that this form might be highly relevant to our decision problem because there we were dealing with a collection of uncertain events having the properties of being exclusive and exhaustive. In that context we denoted them by $\theta_1, \theta_2, \ldots \theta_n$ instead of $E_1, E_2, \ldots E_n$. It is important to remember that in the extended form the events must have the exclusive and exhaustive properties.

As an illustration of the extended form of the theorem, consider the retailer wondering how many items to order from the wholesaler. In our earlier discussion of this example (section 1.8) we saw the decision would depend on the number of items he might sell and we introduced m uncertain events $\theta_1, \theta_2, \ldots \theta_m$; θ_j denoting the uncertain event that he will sell j items. These θ_j are certainly exclusive and we assume them exhaustive. Now suppose the retailer is wondering whether he will be able to pay off his overdraft at the bank: let A denote the event that he will. Then the event will clearly depend on how many items he sells. If he sells a lot he will make a good profit and stands a good chance of clearing his debt; but if he sells few he has little hope of satisfying his bank manager. Consequently he might feel more able to assess $p(A \mid \theta_j)$ for each value of θ_j from θ_1 to θ_m separately than to assess $p(A)$ directly. Thus $p(A \mid \theta_1)$ might be zero or very small. Since he has already assessed $p(\theta_j)$, the likely demand for the items, he now has all the values ready for insertion into the formula and can calculate $p(A)$. For example, suppose $m = 3$ and

$$p(\theta_1) = 0.5,\ p(\theta_2) = 0.3,\ p(\theta_3) = 0.2$$

Then if $p(A \mid \theta_1) = 0.1$, $p(A \mid \theta_2) = 0.3$ and $p(A \mid \theta_3) = 0.8$, we have

$$p(A) = 0.1 \times 0.5 + 0.3 \times 0.3 + 0.8 \times 0.2 = 0.3$$

This is exactly the sort of application we want to make of the extension theorem in the next chapter.

The proof of the theorem is simple. By the multiplication law, $p(A \mid E)p(E)$ is $p(A$ and $E)$; similarly, $p(A \mid \bar{E})p(\bar{E})$ is $p(A$ and $\bar{E})$. Consequently, the right-hand side of equation (3.6) is

$$P(A \text{ and } E) + p(A \text{ and } \bar{E})$$

But the two events 'A and E' and 'A and $\bar{E}$' are exclusive, since E and $\bar{E}$ are, and the addition law may be invoked to show that this sum is the probability of the event {'A and E' or 'A and $\bar{E}$'}. But this event is simply A and the result is proved. The extension to n events is proved similarly.

3.9 THE PUZZLE OF THE THREE PRISONERS

A problem which intrigues many people and also demonstrates the notion of coherence in an interesting way is that of the three prisoners. Alan, Bernard, and Charles are in jail unable to communicate with one another or with anyone besides their respective jailers. Alan knows that two of them are to be executed and the other set free, and after some thinking concludes that he has no reason to think that one of them is more likely to be the lucky one than either of the others. If A denotes the event that Alan will go free, and B and C similarly for Bernard and Charles, this last statement means that

$p(A) = p(B) = p(C) = 1/3$ in Alan's opinion. Alan now says to his jailer 'Since either Bernard or Charles is certain to be executed, you will give me no information about my own chances if you give me the name of one man, Bernard or Charles, who is going to be executed.' Accepting this argument the jailer truthfully says 'Bernard will be executed.' Thereupon Alan feels happier because now either he or Charles will go free and, as before, he has no reason to think it is more likely to be Charles, so his chance is now 1/2, not 1/3, as before. Which argument is correct, the one that convinced the jailer or the latter one?

To analyse the situation we have to compare $p(A) = 1/3$ with $p(A \mid b)$, where b is the jailer's statement that Bernard will be executed. By the first argument $p(A \mid b) = 1/3$ so that b has no effect: by the second, $p(A \mid b) = 1/2$. Now we are told that the jailer is truthful so we know that $p(b \mid B) = 0$ and $p(b \mid C) = 1$; the first statement saying that the jailer will never make his statement if Bernard is going to be freed. It is not clear what value to attach to $p(b \mid A)$, for if Alan is to be freed and therefore both Bernard and Charles to be executed, will the jailer report Bernard's or Charles's execution? So for the moment let us write $p(b \mid A) = p$ and $p(A \mid b) = x$, the quantity of interest: the two values of x considered are 1/3 and 1/2.

By extending the conversation from b to include A, B, and C which are exclusive and exhaustive

$$\begin{aligned} p(b) &= p(b \mid A)p(A) + p(b \mid B)p(B) + p(b \mid C)p(C) \\ &= p/3 + 0 + 1/3 = (1+p)/3 \end{aligned}$$

By the multiplication law

$$p(A \text{ and } b) = p(A \mid b)p(b) = (1+p)x/3$$

but equally

$$p(A \text{ and } b) = p(b \mid A)p(A) = p/3$$

Combining these last two results we have

$$p/3 = (1+p)x/3$$

or simply

$$x = p/(1+p)$$

In the discussion with the jailer, Alan argued that $x = 1/3$. The result just obtained shows that this would imply $p = p(b \mid A) = 1/2$, or that if Alan was to be freed, the jailer would be just as likely to say Bernard as Charles. In feeling happier, Alan was taking $x = 1/2$, which implies $p = p(b \mid A) = 1$, or that in the same circumstances the jailer would always say Bernard.

Hence we see that both assessments, of 1/3 and 1/2, are coherent, but the first implies an even choice on the jailer's part between reporting on the two men to be executed, whilst the second implies that, in Alan's view, he will select Bernard. The essence of the problem, and the reason for its ambiguity, is its failure to state what the jailer will say if Alan is to be freed. If Alan feels elation as a result of the jailer's statement, then that is sensible and coherent provided he realizes it means that Alan is judging that the jailer will report Bernard as the one to be executed in the ambiguous circumstances. This is coherent but probably not what most of us would feel. Most people would opt for $p = 1/2$ and consequently $x = 1/3$ so that the jailer's information is truly irrelevant.

Notice how one can assign seemingly sensible probabilities, $p(A) = 1/3$, $p(A \mid b) = 1/2$, but then use the laws, the rules for coherence, to calculate other probabilities, $p(b \mid A) = 1$, which do not seem sensible. Coherence being inviolate, a revision of the original assignments is needed to make *all* the probabilities agree with one's ideas. Probabilities should not be judged in isolation they should be compared, one with another. We return to this again in section 9.8.

3.10 BAYES' THEOREM

This theorem is named after a person. The Reverend Thomas Bayes was a non-conformist clergyman in eighteenth-century England and was probably the first to state the result explicitly. It is rarely satisfactory to name theorems or ideas after people, since no concept or result is ever entirely due to a single individual. But the reverend gentleman's name is now so firmly associated with this basic result, and with many developments that flow from it, like Bayesian statistics, that the usage must be continued. Descriptive names that might be used are 'inversion theorem' or 'stand-on-your-head' result.

Bayes' theorem

If E and F are any two events then, provided $p(E)$ is not zero,

$$p(F \mid E) = p(E \mid F)p(F)/p(E)$$

Again explicit reference to H is omitted. The proof is trivial. The multiplication law applied to the event 'E and F' says

$$p(E \text{ and } F) = p(E)p(F \mid E)$$

But the event 'E and F' is the same as the event 'F and E', and the law applied to this says

$$p(E \text{ and } F) = p(F)p(E \mid F)$$

From these two results it follows that

$$p(E)p(F \mid E) = p(F)p(E \mid F)$$

and on division by $p(E)$, not zero, we have the theorem.

The importance of the result is that it connects two entirely different probabilities concerning the same two events, namely $p(E \mid F)$ and $p(F \mid E)$. In the former, F is part of the information (remember it is really 'F and H') and E is uncertain, whereas in the latter the roles of E and F are reversed. The English language is sometimes capable of hiding the differences between these two notions. Here is an example of that confusion.

Contrast the two statements:

(1) The death-rate among men is twice that for women;
(2) In the deaths registered last month there were twice as many men as women.

Are these two alternative ways of expressing the same fact, or are they different? In fact they are totally distinct results. To see this, let D denote the event of death, M that of being male, and F that of being female (so that F is the negation of M). Then, using the statistical interpretation of probability referring to repeated events, the statements can be rewritten:

(1) $p(D \mid M) = 2p(D \mid F)$ and
(2) $p(M \mid D) = 2p(F \mid D)$

(1) is a statement about the uncertain event of death, given sex: (2) is a statement about the uncertain event of sex, given death. (2) implies that $p(M \mid D) = 2/3$: (1) does not imply $p(D \mid M) = 2/3$.

Thus the statements are about different things. They could both be true but this would imply that there were equal numbers of men and women: an inequality in the sexes would mean that at least one of (1) and (2) is false. To see this, note that Bayes' theorem says

$$p(D \mid M) = p(M \mid D)p(D)/p(M)$$

and also

$$p(D \mid F) = p(F \mid D)\, p(D)/p(F)$$

Using (1) and (2) in the first of these we have

$$2p(D \mid F) = 2p(F \mid D)p(D)/p(M)$$

The twos cancel and the last two equations together mean $p(M) = p(F)$ and so both are $\frac{1}{2}$.

In contrast, suppose there were twice as many women as men, as can occur in the older age groups; then statement (1) would imply

(2′) in the deaths registered last month there were the same numbers of men as women.

To see this, note that in our notation the supposition says $p(F) = 2p(M)$ and (2′) that $p(M \mid D) = p(F \mid D)$. To prove the latter from the former and (1) use Bayes' theorem,

$$p(M \mid D) = p(D \mid M)p(M)/p(D)$$

With the two facts given this is

$$= 2p(D \mid F)\tfrac{1}{2}p(F)/p(D)$$

$$= p(F \mid D)$$

from a second use of the theorem.

Having shown how different $p(F \mid E)$ and $p(E \mid F)$ are, we can see that Bayes' theorem links them, saying the former is $p(F)/p(E)$ times the latter. In the two probabilities the roles of E and F are inverted, hence the possible name, inversion theorem. In scientific use of Bayes' theorem, F is some general law and E some special case, and we can pass from an uncertainty statement about the special $p(E \mid F)$ (which is often easy) to one about the general law $p(F \mid E)$. The theorem explains how we can make general statements on the basis of special cases.

The result is so important that the whole of Chapter 6 will be devoted to it. For the moment consider one example from a court of law.

3.11 BAYES' THEOREM AND THE LAW

First let us write the theorem in a different way. Combine the original result

$$p(F \mid E) = p(E \mid F)p(F)/p(E)$$

with the same result where F is replaced by its negation $\bar{F}$,

$$p(\bar{F} \mid E) = p(E \mid \bar{F})p(\bar{F})/p(E)$$

dividing the expression on each side of the first equation by the corresponding expression in the other to obtain

$$\frac{p(F \mid E)}{p(\bar{F} \mid E)} = \frac{p(E \mid F)}{p(E \mid \bar{F})}\frac{p(F)}{p(\bar{F})}$$

in which $p(E)$ has disappeared. The ratio of the probability of an event to the probability of its negation under the same circumstances is called the odds and we use the symbol 0, so that $0(F \mid E) = p(F \mid E)/p(\bar{F} \mid E)$. Then the last result reads

$$0(F \mid E) = \frac{p(E \mid F)}{p(E \mid \bar{F})}0(F)$$

This is Bayes' theorem in odds form. (Notice that the way we have defined odds are what a bookmaker would call 'odds on'. He usually refers to 'odds against' as simply 'odds'. In his usage 5–1 means 5–1 against or probability of 1/6. Our odds will be 1/5.)

It is in this form that Bayes' theorem can be used in a court of law. Let E denote some evidence produced in court. Let G denote the uncertain event that the defendant is guilty of the charge brought against him. (G replaces F as a mnemonic.) The unstated H is all the previous evidence and the background knowledge possessed by the judge or jury who is finally to pronounce on the defendant's guilt. Then Bayes' theorem says that the odds of guilt, given the evidence E, are the product of the original odds (without E but, of course, with H) multiplied by $p(E \mid G)/p(E \mid \bar{G})$. In other words, the judge or jury can update their uncertainty of the defendant's guilt as a result of E by multiplying their odds by this factor. The factor is the ratio of the probability of the evidence were the defendant to be guilty to the same probability were he innocent. Notice how the two odds refer to the uncertain event of guilt, but the two probabilities in the multiplying factor refer to the uncertainty of the evidence. All probabilities and odds are those of the judge (or the jury, supposed to act as a single decision-maker). This is the inversion that is at the heart of Bayes' theorem.

Here is an example. A crime of violence has been committed and blood, not the victim's, is found at the scene and presumed to be the assailant's. Evidence E is produced showing that this blood and the defendant's are of the same type. If the defendant were guilty and had bled in the assault then the matching of the types is certain: $p(E \mid G) = 1$. If he were innocent, then someone else committed the crime and the probability of his being of the same blood type as the defendant is f, the frequency of that type. Consequently $p(E \mid \bar{G}) = f$, the ratio $p(E \mid G)/p(E \mid \bar{G})$ is $1/f$, and the odds are increased by multiplying by this quantity which is necessarily greater than one. Notice that the evidence is stronger the rarer is the blood type.

The inversion principle has a useful interpretation in the legal context. Although the judge and jury are concerned with the odds of guilt, the witness, providing the evidence, is not concerned with guilt but with the evidence. The witness is involved with the uncertainty of the evidence on the twin suppositions of guilt and innocence. This is particularly apposite when the witness is a scientist or other expert. Apart from describing E he would also need to say how probable E was in the two situations. In the blood-type case he would provide f (and the value unity in the other possibility). Thus the judge assesses probabilities of guilt; the witness assesses probabilities of the evidence. A judge who believed a witness would use the latter for his own in the theorem.

Generally in assessing the meaning of any piece of evidence, not just in the context of the courtroom, you need to consider how probable the evidence was on one supposition as compared with another. Often only one of the possibilities, guilt in the legal case, is considered but both are needed for a coherent evaluation. An example arose recently when an environmental organisation produced evidence of radioactivity in fish near the discharge from a nuclear power station. But the evidence is inconclusive in establishing the 'guilt' of the power station without knowledge of the radioactivity in fish before the station was discharging.

Finally, notice how in using $p(E \mid G)$ or, in full, $p(E \mid G \text{ and } H)$ it is not known that G is true, that the defendant is guilty, unlike H, which is known. What is being calculated is the probability were he guilty. It is often necessary to use probabilities of this type using the subjunctive mood.

3.12 SIMPSON'S PARADOX

We now discuss a result that is not needed in the mainstream of development of the book, and may be omitted without loss of continuity, but is included because it well illustrates the surprises that are in store when working with probability, and also because it demonstrates how additional data can completely reverse the choice of act.

Our illustration of the paradox concerns a medical trial in which 40 patients were given treatment T and 40 were given a placebo $\bar{T}$. Each patient was subsequently recorded as either having recovered R, or not, $\bar{R}$. The results are given in Table 3.1, together with the calculated recovery rates for the two groups.

It is clear that on this evidence the treatment has been beneficial, increasing the recovery rate over the placebo by 10%. Equating chance with probability, we can say $p(R \mid T) = 0.5$ and $p(R \mid \bar{T}) = 0.4$.

Now the medical workers doing this trial had records of the sex of each patient. There were equal numbers of men and women. The results for the 40 men are given in Table 3.2. Here it is clear that the treatment has been actually harmful for the men, reducing (instead of increasing) the recovery rate by 10%. Since the treatment has been beneficial overall but bad for the men it must surely be good for the women. Their results can be found by subtracting the basic entries in Table 3.2 from the corresponding ones in Table 3.1. Thus 18 treated males recovered in the former out of 20 treated in the latter, whence we deduce two treated females recovered. Proceeding in this way, we have the results in Table 3.3. Our expectation is not realized. The treatment also reduces the recovery rate by 10% in female cases. Here we have a treatment that is bad for men and bad for women but good for all of us! This is the paradox.

Three questions arise: does it matter? how does it arise? can it be avoided? It certainly matters. Given the data for all patients (Table 3.1) the medical profession would certainly recommend the treatment, whereas when the data for sex are included as well the treatment would not be advocated. Material action

Table 3.1. Results for all patients

	R	$\bar{R}$	Total	Recovery rate
T	20	20	40	50%
$\bar{T}$	16	24	40	40%

Table 3.2. Results for male patients

	R	$\bar{R}$	Total	Recovery rate
T	18	12	30	60%
$\bar{T}$	7	3	10	70%

Table 3.3. Results for female patients

	R	$\bar{R}$	Total	Recovery rate
T	2	8	10	20%
$\bar{T}$	9	21	30	30%

is affected by the contrary indications in the paradox. To put it slightly differently: if you had only the overall data (Table 3.1) how would you know that there did not exist some other factor which would completely reverse the results, turning a 10% gain into a 10% loss? In our example the factor was sex: it might have been blood-type, urban or country living, or many other possibilities. So the paradox is really important.

In the medical example the paradox arose for the following reason. The particular treatment under study dealt with an ailment that primarily affected women. (Notice how low their recovery rates are in Table 3.3 compared with those for the men in Table 3.2.) The doctor carrying out the trial did not like the treatment, so he decided to use it mainly on the men, who would mainly recover anyhow, and only rarely use it on women. (Notice that 30 men were treated but only 10 women.) Consequently the treatment was predominantly used with men, having a high recovery rate, whereas the placebo was mostly with the women, having a low rate. As a result the treatment seemed to do better overall. We say that the treatment was mixed up with sex: the technical jargon is confounded with sex.

These considerations show how the paradox can be avoided. Any treatment should be allocated to the patients in such a way that it is not confounded with any other factor, like sex, that might influence the recovery. The reader can easily see this for himself by allocating 20 males and 20 females to the treatment. With the recovery rates for both sexes as in Table 3.2 and 3.3 the results in the modified Table 3.1 will still show a 10% drop as a result of the treatment, and the paradox does not arise. It is reasonably possible to do this with designed experiments like this medical trial. It is almost impossible with survey results where the treatments are not imposed. As an example of the latter, suppose a survey is carried out on the effect of smoking (T) on lung cancer (R). If we merely record results for smokers we could reach Table 3.1 exhibiting a deleterious effect of smoking. But there could exist a genetic factor, replacing sex, that both induces smoking and lung cancer (Table 3.2). What is needed is a designed experiment in which people are selected and then encouraged to smoke or not, so avoiding the confounding.*

We conclude the discussion of the paradox by translating the results into probabilistic terms. Equating chances with probabilities, Table 3.1. says

$$p(R \mid T) = 0.5, \quad p(R \mid \bar{T}) = 0.4$$

* There is no evidence for the existence of the genetic, or other factor, and the causal link between smoking and lung cancer is now established with high probability.

Whereas Tables 3.2 and 3.3 say (using M for male and $\bar{M}$ for female)

$$p(R \mid T, M) = 0.6, \ p(R \mid \bar{T}, M) = 0.7$$

and

$$p(R \mid T, \bar{M}) = 0.2, \ p(R \mid \bar{T}, \bar{M}) = 0.3$$

Extending the conversation from treatment and recovery to include sex,

$$p(R \mid T) = p(R \mid T, M)p(M \mid T) + p(R \mid T, \bar{M})p(\bar{M} \mid T) \tag{3.8}$$

and inserting the numerical values we have

$$0.5 = 0.6 \times p + 0.2 \times (1 - p) \tag{3.9}$$

where $p = p(M \mid T)$. Similarly, replacing T in equation (3.8) by $\bar{T}$

$$0.4 = 0.7 \times q + 0.3 \times (1 - q) \tag{3.10}$$

where $q = p(M \mid \bar{T})$. In our case $p = 0.75$, since 30 out of 40 treated patients were male and $q = 0.25$. Since p is very different from q, the mixing of the treated results in equation (3.9) is quite different from the untreated ones in equation (3.10). Had p and q been the same the mixing would have been the same and the pardox would not have arisen.

3.13 ADDITION LAW AGAIN WITH SCORING RULES

We here clear up a little technicality left unresolved at the end of section 3.4. There we only proved

$$p(E) + p(\bar{E}) = 1 \tag{3.11}$$

rather than the general form of the addition law

$$p(E_1) + p(E_2) = p(E_1 \text{ or } E_2) \tag{3.12}$$

for E_1 and E_2 exclusive. Using the multiplication law, the latter result is easily obtainable from the former. To see this, take the multiplication law, $p(E \mid F) = p(E \text{ and } F)/p(F)$, with $E = E_1$ and $F =$ 'E_1 or E_2', so that 'E and F' is simply E_1. It then reads

$$p(E_1 \mid E_1 \text{ or } E_2) = p(E_1)/p(E_1 \text{ or } E_2)$$

Take this equation with the similar result with E_1 and E_2 interchanged, and add the corresponding sides to obtain

$$p(E_1 \mid E_1 \text{ or } E_2) + p(E_2 \mid E_1 \text{ or } E_2) = [p(E_1) + p(E_2)]/p(E_1 \text{ or } E_2)$$

If E_1 and E_2 are exclusive, given 'E_1 or E_2' means $E_2 = \bar{E}_1$, so that the left-hand side is 1 from equation (3.11). Multiplying by $p(E_1 \text{ or } E_2)$, equation (3.12) is established.

3.14 INDEPENDENCE

In section 3.6 it was noted that the multiplication law in its extension to many events (equation (3.5)) became rather complicated because it involved $p(E_n \mid E_1$ and $E_2 \ldots$ and $E_{n-1})$. There are circumstances, which are fortunately of common occurrence, where the result simplifies. One event E is independent of other events if the probability of E, $p(E \mid H)$, is unaltered by any information concerning the other events. A set of events is independent if the statement in the last sentence is true for each event in the set. In view of the subjective nature of probability (section 2.7) this is a judgement by the subject or decision-maker, and what may be independent for one person may not be so for another.

For two events, E_1 and E_2, independence means (omitting H for clarity)

$$p(E_1 \mid E_2) = p(E_1 \mid \bar{E}_2) = p(E_1)$$

and the corresponding results with E_1 and E_2 interchanged. An obvious example involves two tosses of a coin judged to be fair. If E_i is the event of heads on the ith toss, the above judgements would be usual and the tosses independent. When events are independent, the multiplication law is very simple and, for three events, equation (3.5) becomes

$$p(E_1 \text{ and } E_2 \text{ and } E_3) = p(E_1)p(E_2)p(E_3) \tag{3.13}$$

a simple multiplication of the separate probabilites. It is this form that will be used; for example, in section 6.10.

It is not necessary for independence to be discussed in detail. It must, however, be mentioned that the concept is much subtler than might at first appear. Here is an example included merely to alert the reader to the existence of subtleties. Let a fair die with four faces (triangles) have on the faces 110, 101, 011, 000. Let E_i be the event that when the die is tossed the number on the base has a 1 (not a 0) in the ith place. Thus E_2 occurs with the first and third of the faces in the above list. We leave the reader to observe that each pair of events are independent but that the three together are not independent.

3.15 DUTCH BOOK

The concept of coherence is basic to the ideas of this book. In this section we provide an illustration of the troubles incoherence can bring. In section 3.3 the addition law was proved and, in a special case, gave $p(E) + p(\bar{E}) = 1$. Alternatively expressed, coherence demands that the probability of an event not happening is one minus the probability that it will. Suppose you were incoherent and said $p(E) = 0.2$ and $p(\bar{E}) = 0.7$, adding to less than one. In the language of bookmakers the odds against E from the first statement are 4–1, and you would think it fair to accept, for a stake a, payment of $4a$ to a punter were E to occur. For those unfamiliar with bets, a person would provide a dollars which you would retain unless E occurred, in which case you would return the a together with $4a$ more. Similarly for $p(\bar{E}) = 0.7$, the odds against $\bar{E}$ are 3–7 and for stake b you would think it fair to pay $3b/7$ were $\bar{E}$ to occur. Now suppose both bets are accepted by you with $a = 2$ and $b = 7$. If E occurs you lose $8(=4a)$ from the first and win the stake, 7, from the second. Overall you lose 1. If E does not occur you have the stake, 2, from the first bet but lose 3 $(=3b/7)$ from the second. Overall you lose 1. So you lose 1 whether E occurs or not: the two bets are sure to lose you 1. This is ridiculous, or in our language, incoherent. A combination of bets that is certain to lose money is called, in Britian, a Dutch book. A Dutch book can only be avoided by making the probabilities, and hence the desired odds, coherent.

3.16 DECISION-MAKING

In Chapter 1 a decision problem was structured in terms of a set of decisions $(d_1, d_2, \ldots d_m)$ and a set of uncertain events $(\theta_1, \theta_2, \ldots \theta_n)$. In the last two chapters we have seen how the events can have associated with them probabilites $(p(\theta_1), p(\theta_2), \ldots p(\theta_n))$ all non-negative and adding to 1. The final part of this statement follows from the convexity and addition laws. The probabilities also obey a multiplication law and from all these laws various theorems can be deduced: two have been studied, the extension of the conversation and Bayes. These results will be used in the next chapter to demonstrate that another type of measurement is possible in a decision problem: a measurement of utility. The two measurements, utility and probability, will then be combined to provide a coherent solution to the problem of decision-making under uncertainty.

Before proceeding the reader is advised to do some of the exercises to make sure he has understood the ideas sufficiently to enable him to follows the rest of the argument.

Exercises

3.1. A man has mislaid his diary but remembers he must have left it in a pocket, either of the lounge suit he wore to the office, or of the dinner jacket he had on in the evening. He thinks that the probability of it being in the lounge suit is 3/4. What is the probablity of it being in the dinner jacket?

If it is in the jacket he thinks it is equally likely to be in the left- or right-hand pocket, but if it is in the suit he thinks that, because the right-hand pocket is made differently from the left (in having an inner pocket), it is twice as likely to be in the left as in the right. Use the multiplication law to find the probability that if he looks in the left-hand pocket of the lounge suit he will find the diary. Why is it sensible to look in that particular pocket first?

Extend the conversation to find the probability that it is the left-hand pocket of one or other of the suits.

Calculate the chance of it being in the left-hand pocket of the dinner jacket. Use the addition law and the probabilities calculated in the second and fourth paragraphs of this question to recalculate the probability in the third.

3.2. A gene can either occur in the form A or a and any individual has one of these two forms. (In our language A and a are exclusive and exhaustive: the geneticist uses a

instead of A to denote the opposite of A.) A second gene occurs as B or b. Both forms of each gene are equally likely in that $p(A) = p(B) = 1/2$, but given that one gene is in the capital letter form it is more likely than not that the other is also in the same form; the genes are said to be linked. The same holds for the other form. Suppose that

$$p(A \mid B) = p(a \mid b) = g$$

where g is a number between 1/2 and 1. Use the probability laws to calculate the probability that the two genes are of the same form: in symbols, $p(AB \text{ or } ab)$.

3.3. A part of a factory consists of a number of similar spinning machines attended to by a single operator. The machines work automatically except that every so often a thread breaks and the attendant has to join the loose ends and restart the machine. The probability that exactly one machine is stopped (and therefore being attended to) is 0.40. The probability that two are stopped (one being attended to, the other awaiting attention) is 0.20. The probability of three or more being stopped is 0.20. What is the probability that all machines are working and the operator is idle?

Given that the operator is busy, what is the probability that there is at least one other machine stopped and awaiting attention?

3.4. Continuing with the example of the mountain pass: (as in the text) let E denote the event of the pass being open and A the event of an accident, and suppose $p(E) = 4/5$, $p(A \mid E) = 1/16$ and $p(A \mid \bar{E}) = 1/8$. Let B be the event of arriving late for the meeting and suppose

$$p(B \mid E \text{ and } A) = 1/2 \text{ and } p(B \mid E \text{ and } \bar{A}) = 1/8$$

(In words, the first statement says that if the pass is open but he has an accident, there is a 50% chance of his being late.) Calculate the probability of being late, given that the pass is open, $p(B \mid E)$. (In applying the theorem all events will be conditional on E at least.)

Suppose furthermore

$$p(B \mid \bar{E} \text{ and } A) = 1 \text{ and } p(B \mid \bar{E} \text{ and } \bar{A}) = 3/4$$

Similarly, calculate $p(B \mid \bar{E})$, the chance of being late if the pass is blocked. Hence find the probability of being late, $p(B)$.

(All probabilities are conditional on H, the general circumstances of the problem and the decision to use the car.)

Chapter 4

A Numerical Measure for Consequences

'I take each problem as it comes, I do the sum, I return the answer, and so I act. I act according to a reasoned estimate of what is best.'

Flowering Wilderness, Ch. 3.

4.1 THE CONSEQUENCES OF A DECISION

This chapter parallels Chapter 2. There we tried to show that it was sensible to associate numbers with the uncertain events. Now we show that the consequences of decision-making can also have numbers attached to them, and that these two sets of numbers combine to solve the problem and determine the best decision. We begin with the usual lists $d_1, d_2, \ldots d_m$ of decisions, $\theta_1, \theta_2, \ldots \theta_n$ of uncertain events and also the probabilities $p(\theta_j)$ associated with the latter.

Suppose a particular decision, d_i, say, is selected and the uncertain event, θ_j, occurs. The occurrence of the event will remove all uncertainty from the problem and the action, d_i, will produce a definite result which can be foreseen with certainty. In other words the combination of d_i with θ_j will result in a foreseeable consequence. This consequence will be written C_{ij} or equally (d_i, θ_j). For example, in the mountain pass situation d_i may be the decision to drive the car, θ_j the event that the pass is blocked by snow, when C_{ij} is the consequence that the driver has to turn back.

A second example will illustrate the points to be made more simply. Consider a manufacturer who has just produced an item like a roll of curtain material. He is about to sell it but before he does he has to decide whether or not to inspect it for flaws. There are two decisions; d_1 to inspect, d_2 to sell without inspection. Let us suppose that there are just two uncertain events: θ_1 that the roll is free from flaws, θ_2 that it has at least one flaw. If he inspects and finds the material to be all right (action d_1 with θ_1) the consequence, (d_1, θ_1) or C_{11}, is that he will despatch and sell material free of defects but at the cost of having inspected the material. On the other hand, if he chooses d_2 when θ_2 obtains he will be in the awkward position of selling flawed material and suffering complaints from his customer, although he will have been saved the cost of inspection.

It is convenient to express the situation in the form of a table (Table 4.1). The rows of the table correspond to the decisions, the columns to the uncertain events, and the intersections of rows and columns to consequences. The four possible consequences have been described briefly in the table. Conceptually this form of tabulation is always possible for any decision problem, though with large numbers of decisions or events it would be cumbersome to display it in full. Devices for breaking up a large table will be developed in Chapter 8. For the moment it is easiest to imagine the complete table with m rows, n columns, and $m \times n$ consequences.

In the present example C_{21} is the happiest consequence for the manufacturer, for he will be selling good material without having incurred the cost of inspection. C_{11} is probably the next best, with a satisfied customer only obtained at the cost of a single inspection. C_{12} is next in line: again a satisfied customer but at the cost of having to discard the flawed roll, find a new good one and all the inspection involved. C_{22} is almost certainly the worst of them all because the customer will complain, the roll (as in C_{12}) will have to be replaced, and the only merit is that the manufacturer will, by taking d_2, not have incurred the initial inspection cost.

Hence the four possible consequences can be ranked with the best first:

$$C_{21}, C_{11}, C_{12}, C_{22}$$

Notice that this particular ranking is not inevitable—there are situations where the cost of inspection may be high, but the remedying of the defect so easy once detected, that C_{22} might not be so far down the list. The point we do want to drive home is that there are some preferences amongst the consequences. With the particular ranking just described it is not immediately obvious which decision the manufacturer should take because he does not know whether the material is free from flaws or not. If θ_1 were true, then d_2 is clearly the better, as it is useless to inspect good material, so that C_{21} is preferred to C_{11}. On the other hand, if θ_2 were true then d_1, inspection (followed by subsequent replacement by sound material) is to be preferred and C_{12} is preferable to C_{22}. The manufacturer is in a quandary but, using the arguments of Chapter 2, he can

Table 4.1. Decision table for an inspection problem

	θ_1: Good material	θ_2: Poor material
d_1: Inspect	C_{11}: Satisfied customer, cost of inspection	C_{12}: Find a new roll of good material, plus cost of inspection. A satisfied customer
d_2: No inspection	C_{21}: Satisfied customer	C_{22}: Complaints from the customer, replacement by a new roll

attach probabilities $p(\theta_1)$, $p(\theta_2)$ to these two events. (These add to one, so he need only assign $p(\theta_1)$, say: $p(\theta_2)$ will then equal $1 - p(\theta_1)$.) If $p(\theta_1)$ were near 1, corresponding to his being reasonably sure that the material was sound, then he would choose d_2 and not inspect. If $p(\theta_1)$ were near zero, he would be reasonably certain there was a flaw and d_1, inspection, would be preferred. (A third decision, to overhaul his production process, would probably be better still, but we are working within the framework of two decisions only.) Somewhere along the line as $p(\theta_1)$ decreases from near one to near zero he should presumably change from d_2 to d_1, but where? that is the problem. Obviously where will depend not only on the ranking of the consequences but also on how much better one consequence is than another. For example, if C_{22} gets worse because customer goodwill becomes more important, then $p(\theta_2)$, the chance of poor material, will not have to be so large before inspection begins to be preferred in order to avoid this unpleasant result.

4.2 A NUMERICAL VALUE FOR A CONSEQUENCE

It follows that the next task is to provide something more than just a ranking of the consequences. In order to do this a standard is introduced and coherent comparison with it provides a numerical assessment, just as with the uncertain events. In the case of probability, the standard was balls in an urn. For the consequences, two reference consequences are used; one of these is better than, or at any rate not worse than, any of the consequences in the relevant table; the other is similarly worse than, or at most not better than, all the C_{ij}.

In general let a decision problem be presented in the form of a table with decisions d_i, events θ_j, and consequences C_{ij}. Suppose that any pair of consequences can be compared in the sense that one of them is preferred to the other or alternatively that they are equally desirable. Suppose further that this comparison is coherent in that if a consequence C_1 is preferred to C_2 and C_2 is preferred to C_3, then C_1 is preferred to C_3. That is, we assume coherent comparison of consequences just as in section 2.8 we assumed coherent comparison of uncertain events. (Discussion of this assumption is delayed until later.)

Let C be any consequence which is such that no consequence in the table is preferred to it. Similarly let c be any consequence which is such that no consequence in the table is worse than it. C may be the best, and c the worst, of all the consequences in the table: alternatively they may be new consequences outside those in the table. It follows that any consequence C_{ij} may be compared unfavourably with C and favourably with c.

Now take any consequence C_{ij} and fix on that. Consider the standard of balls in an urn (section 2.5) with U out of the N balls black. Let a ball be drawn at random (section 2.6), if it is black suppose consequence C results, if white, consequence c. Expressed differently, you will receive C with probability $u = U/N$ and c with complementary probability $1 - u$. This will be referred to as 'C with chance u', implying that c has chance $1 - u$. How does 'C with chance u' compare with C_{ij}? If $u = 1$ it is clearly better than C_{ij}, if

$u = 0$ it is worse. Also as the number of black balls U, and hence u, increases the gamble gets better. Hence exactly as with the derivation of probability in section 2.5 there must be a value of U, and hence u, such that you are indifferent between 'C with chance u' and C_{ij}.

It has therefore been shown that associated with any consequence C_{ij} is a unique number u, lying between 0 and 1, such that C_{ij} is just as desirable as a chance u of C and $1 - u$ of c; C and c being a reference pair of highly desirable, and highly undesirable, consequences respectively.

The number associated with C_{ij} will be denoted $u(C_{ij})$, rather than just u, and will be called the *utility* of C_{ij}: hence the choice of letter. Because of the coherence required in the comparison of the consequences it is clear that if C_{ij} is preferred to C_{kl}, then $u(C_{ij})$ will be larger than $u(C_{kl})$; if C_{ij} and C_{kl} are equally desirable then $u(C_{ij})$ will be equal to $u(C_{kl})$; finally if C_{ij} is worse than C_{kl}, then $u(C_{ij})$ will be less than $u(C_{kl})$. We have therefore achieved what we set out to obtain, namely a numerical measure of the desirability of any of the consequences in the decision table; called a utility of the consequence. An extremely important point that we shall repeatedly have occasion to emphasize is that this utility is a probability and therefore obeys the rules of probability. This follows since u is the probability of obtaining the most desirable consequence C.

A decision table contains lists of the decisions as rows and the uncertain events as columns. The columns have associated with them probabilities and the intersection of any row with any column bears a number, the utility of the consequence associated with the decision of that row, and the event of that column. A typical table with three decisions and four events is given in Table 4.2. Table 4.3 is a revised form of Table 4.1, the inspection example, with possible numbers included for the probabilities and utilities, the utilities having been selected to reflect the features of the consequences previously discussed. Thus C_{21} and C_{22} play the roles of C and c respectively and so have utilities of 1.0 and 0.0. C_{11} differs from $C_{21} = C$ by the cost of inspection only, and has been assigned a utility of 0.9. In other words, inspection results in a loss of utility of 0.1. C_{12} involves not only the inspection loss but also the repairing of the flaw or replacement by a new roll, and has been assigned a utility of 0.5. Probabilities have similarly been allocated and it has been supposed that the chance of a roll having a flaw is 0.2. This is perhaps a rather high value

Table 4.2. Decision table: the unnamed entries are utilities

	θ_1	θ_2	θ_3	θ_4
d_1	$u(C_{11})$	$u(C_{12})$	$u(C_{13})$	$u(C_{14})$
d_2	$u(C_{21})$	$u(C_{22})$	$u(C_{23})$	$u(C_{24})$
d_3	$u(C_{31})$	$u(C_{32})$	$u(C_{33})$	$u(C_{34})$
Probabilities	$p(\theta_1)$	$p(\theta_2)$	$p(\theta_3)$	$p(\theta_4)$

Table 4.3. Decision table for the inspection problem with numerical values for the utilities and probabilities

	θ_1: Good	θ_2: Bad
d_1: Inspect	0.9	0.5
d_2: No inspection	1.0	0.0
Probabilities	0.8	0.2

for a realistic problem but it will serve for illustrative purposes and other values will be considered below.

4.3 THE COMBINATION OF PROBABILITIES WITH UTILITIES

Numbers have now been associated with the events and with the consequences. The final stage in the argument is to associate numbers with the decisions, in such a way that the best decision is that with the highest number. It is easy to do this, without invoking any new principle, since both sets of numbers, probabilities and utilities, obey the probability laws and therefore must combine to produce a set of numbers for the decisions in a way prescribed by these laws. To effect this combination consider what will happen if decision d_i is taken. The outcome will depend on the uncertain event and if θ_j occurs the consequence will be C_{ij}, which can be replaced by a chance $u(C_{ij})$ of C, the most desirable consequence. We see therefore that whatever decision is taken, and whatever event obtains, we can think of the result as either C or c. Now if we take d_i, and if θ_j obtains, the probability of obtaining C is $u(C_{ij})$. In the notation of Chapter 3 this reads

$$p(C \mid d_i \text{ and } \theta_j) = u(C_{ij})$$

Since, for the moment, we are only going to consider what happens if d_i is selected, let d_i be omitted from our probability statement in agreement with the convention, mentioned in section 3.2, that if everything depends on some event, there H, here d_i, it is left out of the probability statement, though not forgotten. With this convention

$$p(C \mid \theta_j) = u(C_{ij})$$

The probability of θ_j is also known, $p(\theta_j)$. Apply equation (3.7) here with the events $E_1, E_2, \ldots E_n$ replaced by the events $\theta_1, \theta_2, \ldots \theta_n$. (Remember that it is a condition of the use of the theorem that the events be exclusive and exhaustive and this is certainly true of our θ_i.) Let A of the earlier statement be replaced by the event of obtaining the desirable consequence C. The theorem says

$$p(C) = p(C \mid \theta_1)p(\theta_1) + p(C \mid \theta_2)p(\theta_2) + \ldots + p(C \mid \theta_n)p(\theta_n)$$

It is convenient to introduce the symbol, $\sum$, meaning sum, and to write the last result as

$$p(C) = \sum_{j=1}^{n} p(C \mid \theta_j) p(\theta_j)$$

so that $p(C)$ is the sum, from $j = 1$ to n, of the terms $p(C \mid \theta_j)p(\theta_j)$. Inserting the values for $p(C \mid \theta_j)$ we finally obtain

$$p(C) = \sum_{j=1}^{n} u(C_{ij}) p(\theta_j) \tag{4.1}$$

Remembering that $p(C)$ is really $p(C \mid d_i)$, the probability of C, given d_i, we have arrived at an expression for the probability of obtaining C when action d_i is taken.

Remembering that if C does not result, then the undesirable consequence c necessarily will (with probability $1 - p(C)$), we see that $p(C \mid d_i)$ is a measure of the merit of d_i in the sense that the bigger it is the more desirable d_i is. Consequently the best decision to take is that decision which has the largest value for this probability, and our task is complete. A number, $p(C \mid d_i)$, has been associated with each decision, the larger the number, the better the decision.

The conclusion which results from all the arguments presented up to now can be simply stated as follows. A decision problem can be formulated as lists of decisions and of uncertain events. On the assumptions of coherent comparisons of events, and of consequences, probabilities can be assigned to the events and utilities to the consequences. Each decision can then be assigned a value given by equation (4.1). The best decision is that with the highest value.

This is the major result in the book. Based only on the requirements of coherent comparisons for events and for consequences, it states the inevitability of the procedure indicated.

4.4 EXPECTED UTILITY

Some notation and nomenclature are convenient at this point. The right-hand side of equation (4.1) is called the *expected utility* of d_i and we write it $\bar{u}(d_i)$. (Read 'u bar of d_i'.) Hence, with $C_{ij} = (d_i, \theta_j)$ the consequence of taking d_i when θ_i obtains,

$$\bar{u}(d_i) = \sum_{j=1}^{n} u(d_i, \theta_j) p(\theta_j) \tag{4.2}$$

(The reason for the adjective 'expected' is explained later.) The procedure justified above amounts to choosing that decision with highest expected utility. The maximum value is denoted by

$$\max_i \bar{u}(d_i)$$

or simply $\bar{u}$. (Here max is an obvious abbreviation for maximum, and i indicates that the maximum has to be taken from amongst the utilities obtained for the different values of i; i, of course, ranging from 1 to m.)

The main result can now be stated even more briefly: a decision problem is solved by maximizing expected utility.

In the simple inspection example (Table 4.3), the expected utilities can easily be found. From equation (4.2) and the numerical values given in Table 4.3, we have

$$\bar{u}(d_1) = 0.9 \times 0.8 + 0.5 \times 0.2 = 0.82$$

and

$$\bar{u}(d_2) = 1.0 \times 0.8 + 0.0 \times 0.2 = 0.80$$

Hence $\bar{u}(d_1)$ is the larger, d_1 is the better decision, and $\bar{u}$ is 0.82. Notice that the expected utility for d_1 is only slightly more than that for d_2, so that d_1 is not overwhelmingly better than d_2. This can best be illustrated by seeing what happens if the utilities and/or probabilities change. Suppose the chance of bad material drops from 0.2 to 0.1. We then have $\bar{u}(d_1) = 0.86$, and $\bar{u}(d_2) = 0.90$. Then d_2 is the better decision and its expected utility, $\bar{u}$, has risen from 0.82 to 0.90. In other words, because the chance of bad material has been reduced by half (from 0.2 to 0.1), inspection can be abandoned (change from d_1 to d_2) and the saving is reflected by a rise in expected utility.

Now suppose that with the probabilities at their original values of 0.8 and 0.2 it becomes harder to carry out the inspection so that the utilities associated with d_1 both drop by 0.1 (to 0.8 for θ_1 and 0.4 for θ_2). Then the expected utility for d_1 similarly drops by 0.1 to 0.72. (The reader may verify this by simple calculations.) That for d_2 remains at 0.80 so that inspection has become too expensive to provide the better decision and the maximum expected utility has dropped from 0.82 to 0.80.

4.5 COHERENCE

The recommended procedure for making decisions is to

(1) List the possible decisions ($d_1, d_2, \ldots d_m$);
(2) List the uncertain events ($\theta_1, \theta_2, \ldots \theta_n$);
(3) Assign probabilities to the latter ($p(\theta_1), p(\theta_2), \ldots p(\theta_n)$);
(4) Assign utilities $u(d_i, \theta_j)$ to the consequences (d_i, θ_j);
(5) Choose that decision of maximum expected utility

$$\bar{u}(d_i) = \sum_{j=1}^{n} u(d_i, \theta_j) p(\theta_j)$$

This procedure follows from basic ideas of coherence in judging both uncertain events and consequences. It is the simple message of this book. If we were to adopt the style of a religious text, I would say to you 'Go ye forth and maximise your expected utility'. The remainder of the book investigates ways of carrying out the programme (1)–(5) and the consequences of so doing. In the rest of the present chapter, we discuss the method and, in particular, the idea of coherence on which it is based.

The use of coherence in connection with probability was discussed in section 2.8 and its effect demonstrated in section 3.7. The same principle has been used here to arrive at the concept of utility. It has been supposed that every consequence can be compared with a standard (of 'C with chance u') and that these comparisons fit together, or cohere, in the sense that if C_{ij} is preferred to C_{kl}, which is in turn preferred to C_{mn}, then C_{ij} is preferred to C_{mn}. The coherent comparison of consequences can be defended in the same way as that for events in section 2.8 by noting that violation of the statement in the last sentence by you would amount to admitting you were a perpetual money-making machine. The major objection lies in denying the possibility of comparing all consequences. It is parallel to the objection that not all uncertain events can be compared. Our reply there (section 2.4) was that it seemed impossible to characterize the comparable events: certainly the division into statistical and non-statistical does not work. A similar response is appropriate here: what consequences are not comparable?

Historically, people have found less difficulty in comparing consequences than they have events. For example, the law in Britain in 1941 took £200 as an appropriate award for the loss of expectation of life of an ordinary adult and, apart from the effects of inflation, this figure is adhered to up to the present.

The strongest argument in support of the assumption is the fact that when people have, by force of circumstances, to make comparisons between dissimilar outcomes they do make them because they have to act, and by their actions they reveal their comparisons. For example, town planners sometimes say that the amenity values of open spaces in cities cannot be assessed, yet every day people are making such assessments when they choose between a house near open land and a similar house some distance away at a much lower price. I have yet to hear someone say they could not buy a house because they were unable to evaluate the delights of open land.

It is appropriate at this point to consider the two basic properties (the coherent comparisons of uncertain events and of consequences) together, and see just what it is we are demanding of our decision-making. Really all that is being said is this: suppose that we want our decision-making to be carried out in such a way that the whole process hangs together, it coheres, it is consistent; then it must be assumed of our decision-making that it possesses these two properties, for otherwise it would be wayward and fickle, if not plain ridiculous. We are not asking much of our choice between courses of action. We are not demanding that it be perfect, or in some sense the best: merely that it forms a whole. This is surely a modest assumption. Indeed, to me personally it seems so modest that I can scarcely believe that from it flows such a powerful recommendation as that to maximize expected utility. That is why I have provided proofs that indeed this is so; for without them, why should you proceed in the recommended way? As we shall see, the notion of maximizing expected utility is not obviously sensible, though, in fact, it stands up to all the counter-attacks known to me. Part of the justification for it rests on the two assumptions that have been made. The remainder of the justification is pragmatic, for, as we shall see, it works.

In comparing consequences we have introduced yet more gambling, using 'C with chance u'. The defence of this is exactly as with events (section 2.6): life is a gamble and the argument merely reflects this.

4.6 AN EXAMPLE OF INCOHERENCE

It is not, of course, true that people are coherent: description does not always agree with prescription (section 1.4). But why does the incoherence not get detected, why do people act like perpetual money-making machines? The answer is that it does not often occur to people to check for coherence and, when they do try, it is often quite difficult to demonstrate. As I write this our government in Britain is attacking democracy in the administration of London but defending it in connection with trade unions: expediency, rather than logic or coherence, is the main tool of the politician. Such incoherence is blatant but some is subtle and hard to see. Here is an example which is not too important but has the merit of demonstrating how the probability rules enforce the coherence and does occur in practice. It is related to the adage 'a bird in the hand is worth two in the bush'.

In the preceding discussion a consequence C_{ij} has been compared with 'C with chance p'. (The notation is changed from u to p because we want to think of it as probability, not as a utility.) Let C_{ij} be a substantial reward, say 100 dollars, and let C be a little bigger, say 101 dollars. Then some people would prefer 100 dollars to '101 with chance p' however large p is (though less than one), arguing that they prefer the certain reward of 100 to anything involving even a slight risk (probability $1-p$) of no gain at all, the increase in reward, from 100 to 101, being negligible. Why gamble away a substantial gain (of 100) for the sake of a doubtful, infinitesimal increase (of 1)?

The reply uses the multiplication law of probability (equation 3.5) to establish the incoherence of the case just advanced. Essentially the objector is saying he prefers

(1) A certainty of 100 dollars, to
(2) A chance p of 101 dollars, against $1-p$ of zero gain,

and this, however large p is (provided it is less than one). Presumably therefore he also prefers

(3) A certainty of 101 dollars, to
(4) A chance p of 102 dollars, against $1-p$ of zero gain,

for the same value of p, because the possible loss (101) is greater, whereas the possible gain (1) is the same.

Suppose the alternative (2) is altered so that if the event of chance p occurs the subsequent certainty of 101 dollars, namely (3), is replaced by a chance at 102 dollars, alternative (4). This will make (2) even worse, since (3) is preferred to (4). The effect of the alteration is that he will gain 102 dollars only if both chances occur in his favour: he must first get the 101-dollar option and then win again to increase this to 102. The chance of this, by the multiplication law, is $p \times p$ (written p^2 and read 'p squared') since two independent events have to go in his favour. Consequently, to be coherent he has agreed that (1) is preferred to

(5) A chance p^2 of 102 dollars, against $1 - p^2$ of zero gain.

The argument may be used over again several times to show that he prefers (1) to

(6) A chance p^{100} of 200 dollars, against $1 - p^{100}$ of zero gain.

(Here p^{100} means p multiplied by p, multiplied by p, and so on, 99 times, 100 p's in all; and is read 'p to the hundred'.) Now p is arbitrary so we can make p^{100} as near one as we like. Thus if the odds in (2) are a million to one, those in (6) will be about ten thousand to one. Now most of us would be interested in a gamble (6) which had odds of ten thousand to one of increasing a gain from 100 to 200 dollars, for it amounts to doubling one's holding with scarcely any risk. If ten thousand to one is not enough then some long odds will be, and these we can get by raising p. Essentially we like the gamble (6) because most of us can sharply distinguish between 100 and 200 dollars even if we are muzzy about any difference between 100 and 101, and if 200 is not sufficient the argument can be used enough times to make it enough: thus 1000 dollars can be had with probability p^{900}.

Consequently the original argument is incoherent. The preferences of (1) to (2) and (3) to (4) imply a preference of (1) to (6) which contradicts the preference of (6) to (1) just stated. The way out of the difficulty is to admit that (2) is preferred to (1) for p sufficiently near 1, and to admit that the requirement of coherence demands this reversal of choice. As we have said, we attach little importance to this argument as such: it is included because it illustrates the way in which the concept of coherence implies an organization in one's way of thinking.

4.7 EXPECTED VALUES

We turn next to an examination of the term 'expected' that was used in describing the quantity $\bar{u}(d_i)$ calculated in equation (4.2). The reason for its use is historical. Consider a gamble in which one stands to gain a dollar with probability p or lose a dollar with probability $q = 1 - p$. If $p = 1/2$ the gamble is often said to be fair. This simple type of gamble occurs in many elementary games of chance, though usually these are sub-fair (that is, p is a little less than

1/2) so that the player has the impression that it is fair but the casino, against whom he is operating, has just that little edge that enables it to make a profit when operating against a large number of players, over a long period of time. Now consider what will happen if this is played, say, a million times. That is, a ball is drawn at random from an urn with a proportion p of black balls; if black the dollar is won, if white, lost; the ball is returned to the urn and the procedure repeated a million times. Then one would expect (note the word) that a proportion about equal to p of these million plays would result in a win. If this was not so one would doubt the random mechanism. Indeed one can prove, from our assumptions, the truth of a more precise form of this statement: it is called a law of large numbers but its details need not concern us here. If the expected happens then one would win $M \times p$ dollars (where M denotes a million) and lose $M \times q$ dollars, and hence gain

$$Mp - Mq$$

dollars. This is the expected gain after a large number, M, of plays. The expected gain per play is therefore this quantity divided by M, namely

$$p - q$$

Similarly if one gained u_1 for a win and lost u_2 otherwise, in a game with the same chances, the expected value of a play of the game would be

$$u_1 \times p - u_2 \times q$$

More generally if a game has probabilities $p_1, p_2, \ldots p_n$ of winning amounts $u_1, u_2, \ldots u_n$ respectively (some of the u's may be negative) then, on the basis of a large number of plays, the expected value of a single play is

$$u_1p_1 + u_2p_2 + \ldots + u_np_n$$

The form of this expression is the sum of a number, n, of terms, each of which is a value times the probability of obtaining that value. Generally, if a quantity can take on values $u_1, u_2, \ldots u_n$ with probabilities $p_1, p_2, \ldots p_n$ (these probabilities referring to exclusive and exhaustive possibilities and therefore adding to 1) the *expected value* is defined to be

$$\bar{u} = \sum_{j=1}^{n} u_j p_j$$

It represents what one can expect to obtain from $u_1, u_2, \ldots u_n$ occurring with probabilities $p_1, p_2, \ldots p_n$. An alternative term is *expectation*.

In the situation considered just before equation (4.2) the utilities $u(d_i, \theta_1)$, $u(d_i, \theta_2)$, $\ldots$ $u(d_i, \theta_n)$ could arise with probabilities $p(\theta_1)$, $p(\theta_2)$, $\ldots$ $p(\theta_n)$, and the expression in equation (4.2) is exactly of the form just described. Consequently it is the expected utility, given d_i. What we have done here is to take over a concept from situations which can be repeated indefinitely and to apply it to a single occasion, namely the choice of d_i. Because the same type of mathematical expression—namely a sum of values times probabilities of these values—has arisen the same word has been used. (It is somewhat like using the word 'horse-power' to describe the power supplied by an internal combustion engine.) Notice that the justification given before equation (4.2) for using expected utility has nothing to do with the notion of indefinite repetition but comes directly from the extension of the conversation.

As we have said before, we are concerned with a single choice of action and neither our notion of probability nor our use of expectation depend on any idea that we might repeat our decision-making. We are concerned with doing something now, on this single occasion. Much of the literature on probability, and almost all that on the concept of expectation, has been devoted to these ideas in the context of repetition. Now whilst it is true that the evaluation of probabilities (and hence of expectations) often uses repetition—as when we determine a chance by consideration of what happened on previous similar occasions—almost invariably we want to use a probability or expected value in reference to a *single* realization. The statistical form of probability does not permit this. For example, in stating that the expected gain on a play of a game is 5 dollars, the usual interpretation is that if it were to be repeated a large number of times, the return per play would be 5 dollars. The reader should be clear that this is not what we mean.

4.8 PRACTICAL IMPLEMENTATION

The reader who has followed our argument and is convinced by it may still be unimpressed because of the difficulties of implementation: how are the probabilities and utilities to be obtained? It is all very well to have a neat, logical theory but if it proves impossible to use it the theory is not of interest. Or as an eminent industrial statistician put part of the objection succinctly: 'You don't catch me guessing at probabilities'.

Our answer to this objection is that if you are convinced by the logic then ways should be found to implement it. At the moment very little effort has been put into the assessment of probabilities or utilities because people have not felt the need to do so whilst the inevitability of the maximum expected utility has not been appreciated. But once it has, then some effort seems called for. There is an argument by analogy. The basic ideas discussed in this book were essentially discovered by Frank Ramsey, who worked in Cambridge in the 1920s. To my mind Ramsey's discoveries in the twentieth century are as important to mankind as Newton's made in the same city in the seventeenth. Newton discovered the laws of mechanics, Ramsey the laws of human action. Whether this comparison is fair or not, imagine a similar reaction to Newton's laws as has just been mentioned for Ramsey's. What is the use of all this theory of masses, accelerations, distances if we cannot measure them? Such an objection would be silly. Faced with Newton's cogent reasoning, scientists proceeded to invent ways of measuring the quantities required. They designed and built apparatus: they used the theory of errors to relate their imperfect values to the ideals of the theory. We should do the same as these scientists. We need 'theodolites' to measure probabilities: we need an error calculus to relate guesses to the concepts of Ramsey. This will be discussed in Chapter 9, but in a sense the rest of the book is all about it because methods are developed from the basic principles to deal with a variety of situations, and such methods are a necessary adjunct to implementation.

Another response to the objection that the theory cannot be used is that people already are making decisions and all we are doing is to suggest that they do it in a simpler way which logically must be better, because of coherence. (One possibility is that people are naturally coherent and do not need Ramsey's ideas. There is ample empirical evidence to show this is false.) Let us illustrate this with reference to the inspection example that began this chapter (Table 4.1). People at the moment are making decisions about whether or not to inspect. In some situations manuals of quite complicated procedures are available to aid them. In others, managers proceed by intuition and experience without formal aids. Now consider what has to be brought to bear on the problem. There is the possibility of defective material, the cost of inspection, the inconvenience of a dissatisfied customer: all these factors have to be allowed for. Can a person really do this? Yet he has to do it—a decision is forced upon him. What we are asking him to do is to focus on separate aspects of the problem, consider them, and then we provide ways of combining them. What possibility is there of defective material? Assess a probability. How expensive is inspection, and how serious is loss of customer goodwill? Assess utilities. Our questions are much simpler than the question of whether to inspect because they isolate elements of the problem and then we provide ways of putting answers together, through maximization of expected utility.

The truth of the matter is that people dislike simple problems: they like to take refuge in complicated ones where the inadequacies of their procedures are difficult to challenge because of the obscurity generated by complication. As has been said, 'Practical decision-makers instinctively want to avoid the rather awful clarity that surrounds a really simple decision'. The reply to the accusation of guessing at probabilities and utilities is simply that if you can't do simple problems, how can you do complicated ones?

To which a reply is: there are complicated problems that people can solve without being able to solve the simple ones that underlie them—for example, riding a bicycle. Is decision-making like riding a bicycle? I think not. People ride bicycles by repeated practice until, one day, it comes to them; they can do it, they never forget, and they don't know how they do it. None of these notions easily carry over to decision-making.

Before we leave this issue of implementation it should be pointed out that where procedures for inspection are laid down rather precisely, as in the assessment of military equipment or in the safeguarding of weights and measures, these are all incoherent. The full discussion of this is outside the technical scope of this book, but, for example, the European Economic Commmission has recently introduced weights and measures legislation that uses incoherent notions. Someone who was in part responsible for the rules told me that coherent ideas would not be understood.

Before proceeding to explore further the concept of maximization of expected utility, there are two small matters that need to be discussed.

4.9 DEPENDENCE OF PROBABILITIES ON DECISIONS

Our argument has used probabilities $p(\theta_j)$ that do not depend on the decision adopted. In practice they can, and we should write $p(\theta_j \mid d_i)$ for the probability of θ_j given d_i. An example of this is first given and then it is shown that the generalization does not materially influence the procedure. We previously used the simpler $p(\theta_j)$ to avoid complications.

Table 4.4 provides an example where the decision problem is a choice between two systems. One (d_1) is cheaper than the other (d_2). The uncertain events concern the possibility of an accident (θ_1) or not (θ_2). The consequence (d_i, θ_j), as before, is written C_{ij}: its utility $u(C_{ij})$. Because of the expense, $u(C_{12}) > u(C_{22})$ and $u(C_{11}) > u(C_{21})$. Due to the accident, $u(C_{11}) < u(C_{12})$ and $u(C_{21}) < u(C_{22})$. On these considerations d_1 is the preferred choice. The novelty is that d_2 is the safer system, the probability of an accident being less with it than with d_1. In symbols, $p(\theta_1 \mid d_2) < p(\theta_1 \mid d_1)$.

To see that the complication does not affect the method, reconsider the argument in section 4.3. We there had $p(C \mid d_i \text{ and } \theta_j)$ but omitted the d_i for clarity. But it could, and perhaps should, have been retained. Then in extending the conversation from C to include the θ's we should have had to combine $p(C \mid d_i \text{ and } \theta_j)$ with $p(\theta_j \mid d_i)$, not just $p(\theta_j)$. We would then have had

$$p(C \mid d_i) = \sum_{j=1}^{n} p(C \mid d_i \text{ and } \theta_j) p(\theta_j \mid d_i)$$

which, since $p(C \mid d_i \text{ and } \theta_j) = u(C_{ij})$, is still the expected utility of d_i. In other words, a different set of probabilities for each row of the decision table does not change the argument that it is expected utility that needs to be maximized.

Table 4.4

	θ_1: Accident	θ_2: No accident
d_1: Cheap system	$u(C_{11})$	$u(C_{12})$
d_2: Expensive system	$u(C_{21})$	$u(C_{22})$

4.10 THE STANDARD FOR UTILITY

Our argument used two reference consequences, C and c. We have to show that the same result would have been obtained had other references been used. The demonstration can be omitted without loss of continuity but does, as a by-product, demonstrate a property of utility, namely that it is unaffected by a linear change, and also uses a simple tree of the type to be studied in detail in Chapter 8.

Let **C** and **c** be two other reference consequences. We suppose **C** is better than C and **c** worse than c. This will ensure **C** is better and **c** worse than all consequences C_{ij}.

Now C_{ij} was equivalent to 'C with chance u'. This is expressed in Figure 4.1 by C_{ij} being associated with two branches (the figure is called a tree) leading to C and c respectively, with labels u and $1-u$ corresponding to the probabilities of the branches. Since c is intermediate between **C** and **c**, it is associated with two further branches leading to these values with probabilities p and $1-p$ for some p. Similarly C has branches to **C** and **c** with probabilities $p+s$ and $1-p-s$. Since C is preferred to c, $p+s$ exceeds p and s is positive. So C_{ij} can be replaced by a pair of branches, and each member of the pair can be replaced by other pairs. Consequently C_{ij} can be replaced by four branches. Two of these lead to **C**, two to **c**. Hence C_{ij} is replaced by a gamble on **C** and **c**, rather than C and c. What are the new probabilities that replace the original u and $1-u$? By the multiplication law the top route to **C**, going by C, has probability $u(p+s)$. Similarly, the lower route by c has probability $(1-u)p$. By the addition law the total probability for **C** is

$$u(p+s)+(1-u)p = up+us+p-up = us+p$$

and the reader can easily verify that the total probability for **c** is 1 minus this. Hence the original utility u for C_{ij} has become $us+p$ when (C, c) is replaced by $(\mathbf{C}, \mathbf{c})$. Here p and s are two positive numbers that do not depend on C_{ij}: that is, they would be the same whatever consequence we started with.

Replacing u by its full description $u(d_i, \theta_j)$, the new expected utility is

$$\sum_{j=1}^{n} [u(d_i, \theta_j)s + p]p(\theta_j)$$

$$= s\bar{u}(d_i) + p$$

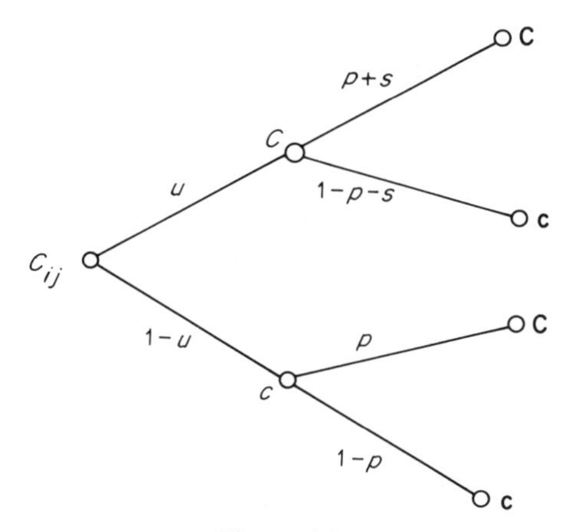

Figure 4.1

in terms of the original one. Since p and s do not depend on d_i, maximization of this new quantity is equivalent to maximization of the original $\bar{u}(d_i)$. Hence the choice of reference consequences is immaterial.

Notice that the utilities (and expected utilities) are all scaled by s and have p added to them as a result of the reference change. We see therefore that the origin and scale of utility does not matter. In the next chapter, when the utility of money is discussed, we shall, like probability, confine utility to values between 0 and 1. But we could have any range, say 32 to 212 by adding 32 and scaling by 180. It will be convenient to use different origins and scales for utilities in different problems, just as with length, light-years are used in astronomy and microns in atomic physics. The unit will be called a utile. Thus we shall speak of a utility of 0.5 utiles. In the next chapter utility is studied in more detail for special types of consequences involving money.

4.11 RISK

There is one term that is frequently used in scientific discussion of decision-making and that is risk, or its derivatives, like risky. It will not be used in this book and the omission had perhaps best be explained. Of course, the word in its usual English sense is perfectly appropriate and it would be sensible to call the situations we are discussing 'risky' because of the uncertainty present, as we do in section 5.12, for example. It is the technical term 'risk' that will not be used. The expected utility, equation (4.2), is often called the risk. We prefer 'expected utility' because it emphasizes what it is, just a form of utility, rather than a new phenomenon, as 'risk' suggests.

The word is often used to make a distinction which is unnecessary for us. The distinction is between those cases where the probabilities are known and those where they are unknown. For example, a decision problem in which the uncertainty enters only through the fall of fair dice would be called risky, whereas one where the uncertainty concerned a government's re-election would not. The reason being that the dice probabilities are well understood and agreed by all (though some might dispute the fairness of the dice) whereas people disagree about the election possibilities. Clearly there is a distinction but to us it is unimportant. To us the distinction is only one of degree, it being harder to assess one than another. In either case the unique probability is the number to be used, together with utility, in maximizing expected utility. The term is therefore not used but our reader should be aware of it because of its frequent occurrence. There is even a journal with the word in its title. The term 'risk-averse' will be introduced in section 5.9 and 'risk premium' in section 5.12.

Exercises

4.1. In each of the following situations one has a game with various possible prizes at stated probabilities. How much would you expect to win at a single play of each of these games?

(i) Prizes:	−1	0	+1				
Probabilities:	0.2	0.5	0.3				
(ii) Prizes:	1	2	4	8	16	32	64
Probabilities:	1/2	1/4	1/8	1/16	1/32	1/64	1/64
(iii) Prizes:	−100	+5					
Probabilities:	0.01	0.99					

4.2. Determine the decision of highest expected utility in each of the following tables (*pr* denotes probability)

(i)

	θ_1	θ_2
d_1	0.0	0.7
d_2	1.0	0.2
pr	0.3	0.7

(ii)

	θ_1	θ_2
d_1	0.4	0.8
d_2	1.0	0.0
pr	0.6	0.4

(iii)

	θ_1	θ_2	θ_3
d_1	1.0	0.5	0.3
d_2	0.4	1.0	0.6
d_3	0.0	0.8	1.0
pr	0.4	0.2	0.4

4.3. Why is the following decision problem capable of simple solution without thinking about the probabilities, and what is the solution?

	θ_1	θ_2
d_1	0.0	0.4
d_2	1.0	0.6

4.4. A decision-maker had determined his utilities and was about to consider the relevant probability p:

	θ_1	θ_2
d_1	0.6	0.8
d_2	1.0	0.0
pr	p	$1-p$

Find what value p would have to be for the expected utilities of d_1 and d_2 to be equal. Hence show that if a quick think convinces him that p is less than 1/2 he need not consider p any further but should reach a decision. Which decision?

4.5. A decision-maker had obtained the following table:

	θ_1	θ_2
d_1	u	0.75
d_2	1.0	0.0
pr	0.6	0.4

but had difficulty in determining the utility of the consequence C_{11}. Show that a critical value for u is 0.5 but that if u is not near this value then it is easy to make a decision without knowing its exact value. What decisions would you make for $u = 0.1$ and $u = 0.9$?

4.6. (The following is an attempt to imitate, in a very simple situation, some of the features of certain card games.) A pack of 9 cards contains 2 aces, 3 kings and 4 jacks. A card is drawn at random, the player looks at it and can either (d_1) call for a second card from the reduced pack of 8 or (d_2) decline to play. If he chooses d_2 then there is no payment. If he selects d_1 then the rewards are as follows: 2 aces or 2 kings give a prize of 2 dollars, 2 jacks or an ace and a king give a prize of 1 dollar, other combinations lose a dollar. For each possible first card find the optimum choice between d_1 and d_2 and the expected value of the game. If the whole game requires an admission fee from the player, what fee would make the game fair?

4.7. At British parliamentary election a firm of bookmakers was offering odds of 4–7 against Labour and 5–4 against the Conservatives. Equating utility with money (a point to be discussed in detail in the next chapter) show that one should bet unless one's personal assessment of Labour's chances of winning lies between 0.56 and 0.64.

4.8. Consider Exercise 1.4. Suppose the treatments cost 3, 2 and 1 units for t_1, t_2 and t_3 respectively, and suppose $p(\theta_1) = 0.2$, $p(\theta_2) = 0.5$, $p(\theta_3) = 0.2$, $p(\theta_4) = 0.1$. On the basis of cost alone what is the doctor's optimum decision?

Chapter 5

The Utility of Money

'James wiped his napkin all over his mouth.
"You don't know the value of money," he said, avoiding her eye.'

The Man of Property, Ch. 3.

5.1 THE USE OF ASSETS, RATHER THAN GAINS OR LOSSES

Utility is a number measuring the attractiveness of a consequence—the higher the utility, the more desirable the consequence—the measurement being made on a probability scale. It is sometimes difficult to attach a number to a consequence because the relevant features are not naturally quantifiable. The pleasures to be derived from a walk in the country, a visit to a theatre, or winning an argument with a colleague are aesthetic and psychological and not immediately expressible in numerical terms. (In passing, note that one buys a theatre ticket or a train ticket to the country, so that there is some measured element in those situations.) Therefore it will be simpler if we start our study of utility by considering cases where there is already a numerical value present. The situations to be considered are those in which the consequences are entirely monetary. Examples are bets in which one stands to win or lose prescribed sums of money, investments on the stock exchange, insurance against loss of tangible assets, or the many circumstances in industrial technology where the outcomes of the choice of any particular design can be expressed in terms of costs and rewards. Almost all consequences have some monetary element (for example, the visit to the theatre cited above) but in the present chapter our concern is with those that are *entirely* monetary and do not have aesthetic, psychological, or moral overtones.

Consider someone contemplating the investment in a stock of 500 dollars for a period of three months. At the end of that time the stock may be worth more or less than the original sum spent on purchasing it. Suppose, for simplicity, that at the end of three months the stock will either have appreciated to 600 dollars or have dropped to 400. In practice there are many more than just these two possibilities, but the simpler situation will capture the essence of the argument whilst avoiding arithmetic complexities. Hence there are two uncertain events: θ_1, increase to 600; θ_2, decrease to 400. With two decisions, d_1 to invest and d_2 to keep the 500 dollars in the bank, we have a simple decision problem in the standard form already discussed. It is presented in the usual tabular form in Table 5.1, except that the entries in the body of the table are the monetary outcomes now about to be described, and not the utilities.

Consider any one of the consequences, say investment followed by a fall in price (d_1, θ_2). The important feature is a loss of 100 dollars compared with the result of not investing. This is a comparison of two consequences, not a description of one. In order to obtain the latter suppose that the investor had capital of amount C at the time he bought the stock, then (d_1, θ_2) can be described by saying his capital is now $C - 100$, assuming no other factors have affected it in the intervening period. The correct monetary entries are those given in Table 5.1, and it is to those that utilities must be attached, not to the losses or gains of 100 dollars, for these do not describe consequences but only differences between consequences. For any capital sum, x, the utility of x will be written $u(x)$. In the example we require $u(C-100)$, $u(C)$, and $u(C+100)$. One immediate result of these considerations is that the decision whether or not to invest may depend on C, the capital at the time the decision is made, and this is surely a realistic conclusion.

Throughout this chapter we shall therefore be considering monetary consequences, where the sum of money involved describes the total assets of the decision-maker if that consequence results. The capital concerned is the total realizable capital and not just the fluid surplus capital, as can be seen by recognizing that a decision whether to invest or not may depend on the former and not merely on the latter. The security that comes from the ownership of property, which can be mortgaged if necessary, affects one's investment policy. So the sums of money that we are discussing are always non-negative and it is not sensible to talk of negative assets. The law recognizes this when fines are levelled against offenders or maintenance orders imposed on defecting husbands, and no attempt is made to extract more than the litigants possess.

Our task, therefore, is to discuss the form of the utility function $u(x)$ describing the relationship between utility and total monetary assets, x. Before doing this we introduce an example designed to illustrate two points: first, the need for such a function; and second, but perhaps more important, the way in which the coherence principles that have been advocated work in practice.

Table 5.1. The entries are sums of money

	θ_1: Stock appreciates	θ_2: Stock depreciates
d_1: Invest	$C+100$	$C-100$
d_2: Leave in bank	C	C

5.2 INCOHERENCE

The example concerns four bets listed as follows:

Bet	Lose	Win
I	10	10
II	10	20
III	20	10
IV	20	20

Thus bet I, if accepted, will either lose you 10 dollars or win you 10. (Dependent on the reader's interpretation of our dollar, and also on his assets, he may wish to scale these values up or down—for example, by multiplying by 10—in order to increase the interest of the bets.) So far no probabilities have been mentioned. The reader, before proceeding further, is asked to state for each bet the *least* value of p, the probability of winning, that will lead him to accept the bet. The bet at this value of p must be as attractive as declining the bet and preserving the *status quo*, since a smaller value would mean declining the bet and a larger value would make it more worthwhile.

In carrying out the requested task the reader has solved four separate decision problems, in each case the decisions being to accept or to refuse a bet. Thus, if p_1 is his stated value for the first bet, he has decided to accept if p exceeds p_1 and to refuse otherwise. Let us now see how the four solutions fit together, or cohere. Experience with the example on subjects unfamiliar with utility concepts has shown a great variety of reactions, some rather ridiculous and others apparently sensible but still incoherent. To illustrate consider assessments in the latter class with $p_1 = 0.6$, $p_2 = 0.5$, $p_3 = 0.8$, and $p_4 = 0.6$. These are reasonable since II is clearly the most favourable bet and would have the least value of p, whereas III is the worst and would require the largest probability, the others being intermediate and differing only in the equal sums to be won or lost. Nevertheless we proceed to show that these values are incoherent.

(Some readers will have chosen $p_1 = 1/2$, $p_2 = 1/3$, $p_3 = 2/3$, and $p_4 = 1/2$; that is, those values that make the bets *monetarily* fair. For example, with bet III, $-20(1 - p_3) + 10p_3 = 0$ when $p_3 = 2/3$. Such decisions are, of course, coherent and the utility of x dollars is equal to x dollars over the range of x-values involved in the bets. These readers may care to change their interpretation of the dollar to make it a larger amount, or alternatively to multiply all the amounts by, say, 100. If a sufficiently large factor is chosen the above probabilities will cease to be reasonable. Generally in this chapter we are considering important gambles where the rewards and/or losses are large in comparison with the assets. With smaller gambles one can work with values that are monetarily fair and treat utility as proportional to money.)

In assigning these four values, the decision-maker has admitted that the four bets at these probabilities are equally desirable, because they are all equivalent in his mind to the *status quo*. Now consider what is called a *mixture* of bets. This is a situation in which one of the four bets is selected by chance with probabilities which we denote by a_1, a_2, a_3, a_4. For example, with $a_1 = a_2 = 1/2$, $a_3 = a_4 = 0$ a fair coin might be tossed and if it falls heads the bet I is selected whereas tails results in bet II. Then since these two bets (at the selected probabilities) are equivalent to the *status quo*, the same must be true of the mixture. In the numerical example take

$$a_1 = 0.3, a_2 = 0.4, a_3 = 0.3, a_4 = 0.0$$

and call this, mixture A. Then bet I will be used with chance 0.3, and if it is, 10 dollars will be won with chance 0.6. By the mutliplication law the chance of bet I being used and resulting in a win is $0.3 \times 0.6 = 0.18$. 10 dollars can also be won with bet III, the same argument giving a chance of $0.3 \times 0.8 = 0.24$. These are the only two ways a win of 10 dollars can be obtained, so by the addition law the chance of such a win is $0.18 + 0.24 = 0.42$. Proceeding similarly with the other sums, the mixture with the above probabilities leads to the probabilities given the first row of the following table

Mixture	Lose 20	Lose 10	Win 10	Win 20
A	0.06	0.32	0.42	0.20
B	0.06	0.355	0.42	0.165

The second line is similarly obtained from a mixture with

$$a_1 = 0.7,\ a_2 = 0.15,\ a_3 = 0.0,\ a_4 = 0.15$$

called mixture B.

Now both mixtures are equivalent to the *status quo* and therefore are themselves equivalent, in the sense that the decision-maker, if coherent, should have no preferences between A and B. This is plainly nonsense, because the mixtures have the same chances of winning 10 or losing 20 dollars, the only difference between them being that A has a higher chance of winning 20 dollars and a lower chance of losing 10, and as a result is preferred to B. Thus in assigning the four probabilities $p_1 = 0.6$, etc. the decision-maker has been incoherent.

The example has demonstrated our second point, namely the manner in which coherence operates. The avoidance of incoherence is achieved through a utility function, but a demonstration that is so must be delayed until the function has been studied in more detail. Notice that in its combination of bets leading to absurdity the ideas of this section are closely related to those of a Dutch book discussed in section 3.15. The treatment here takes full account of the utility structure: that in section 3.15 effectively supposed utility to be the same as money. (Some readers may wish to know how the mixtures A and B were found. One way is by trial and error. A more efficient procedure is to express the problem as one in linear programming. To discuss this would take us too far away from the central topic of this book but the reader familiar with linear programming may like to carry out the exercise. We return to the discussion of a utility function for money.)

5.3 UTILITY IS INCREASING AND BOUNDED

The first obvious remark is that an increase in x causes an increase in utility, so that $u(x)$ increases with x. Most of us would prefer the larger of two sums of money, and since utility measures the desirability of the money, the larger sum must have the larger utility. If this were not so then we should find people literally throwing money away. Experience shows that people do not do this. They may give money away but this is usually associated in their minds with a gain at least in utility for society and the consequences are not entirely monetary.

A second reasonable feature of $u(x)$ may be obtained by considering very large values of x. This is a little harder to think about because we are so unfamiliar with really substantial assets. However, suppose you contemplate a consequence with which is associated such a value, say a win of a million dollars; or if that is not large enough (as it may not be if the decision-maker is an industrial corporation) some really big value. Then you would find it quite hard to distinguish between an x of one million and an x of two millions. Admittedly the latter is better, but the former would enable you to do all those wonderful things you have wanted to do for so long, and buy all those marvellous extravagances that you never thought you would be able to have, so that a further million dollars on top would only gild an already very attractive lily. To put it differently, there would come a point where the extra capital would cease to excite you. It therefore seems natural to suppose that utility does not increase without limit as x does, but that utility is bounded by some upper limit as x increases. This upper limit need never be attained for any value of x but values of x can be found whose utility is as near to the limit as is desired. A glance at Figure 5.1, or at the two figures at the end of the book, will demonstrate the point.

Since $u(x)$ increases with x and x cannot fall below zero, the utility of zero, $u(0)$, must be a lower bound for utility. Similarly we have just seen there is an upper limit. It will agree with the discussion in section 4.10 if we assign a utility of zero to x being zero (that is, put $u(0) = 0$), and utility of 1 to the upper limit. The worst possible consequence, c, will correspond to $x = 0$. The best possible consequence, C, is this upper limit which is just slightly beyond any attainable value of x and for a large value of x, $u(x)$ is almost 1. The outcome of these considerations is that $u(x)$ is reasonably an increasing function of x, with $u(0) = 0$ and an upper limit, as x increases, of 1. Such functions are illustrated in the figures in this chapter and at the end of the book. We saw in section 4.10 that it does not affect decision-making if the utility values have a constant added to them, or if they are all multiplied by a constant.

The use of 0 and 1 as the lower and upper limits enables us to employ the method of obtaining utilities described in the previous chapter, namely to replace any sum x by a gamble with some chance of a utility of one and a complementary chance of zero, the chance being equated to the utility, $u(x)$. This method is unsatisfactory here if only because we find it so hard to think of these extreme utilities of 1 and 0, representing perfection and disaster, respectively. As has been emphasized before, that device was used to establish the existence of utilities, for which purpose it is perhaps the simplest. To explore the values of utilities in most situations it is preferable to employ other checks on coherence which are simpler to gauge, just as it is better to use theodolites rather than rulers to measure distances of the order of miles.

5.4 DIMINISHING MARGINAL UTILITY

Consider again the investor contemplating purchasing stock. Suppose that he does so and that the stock appreciates, so that he gains 100 dollars, his initial capital growing from C to $C+100$. Then his gain in utility is $u(C+100)-u(C)$, and the difference measures the satisfaction he obtains from the additional 100 dollars. Now for most of us this satisfaction will depend on C, the initial capital. If C is also 100, the gain represents a doubling of capital and the satisfaction is presumably rather high. On the other hand, if C is a million then the gain is infinitesimal in comparison and there is little satisfaction to be had. These two extremes illustrate the phenomenon that, for many of us under typical conditions, the increase in utility derived from the increase of 100 dollars in capital is smaller the larger the initial capital. We express the idea more generally and more mathematically.

For a fixed gain a in monetary capital, the increase in utility

$$u(x+a)-u(x)$$

is a diminishing function of the initial monetary capital, x. (Here a is positive, as also is x by a previous assumption.)

This is often referred to as the principle of the diminishing *marginal utility* of money, marginal utility being the term used for the increase in utility due to an increase in capital, as distinct from the utility of capital. Before discussing this any further it is necessary to make another observation.

5.5 A DISTINCTION BETWEEN NORMATIVE AND PRESCRIPTIVE VIEWS

In the first four chapters of this book it has been shown that if certain assumptions are made then certain results necessarily follow; there is no disputing the truth of the results granted the premises. Furthermore we have argued strongly in favour of these premises. The properties of utility that are being discussed in the present chapter do not have this inevitability and they do not follow from the assumptions. We merely put them forward as reasonable attributes of a decision-maker's utility function. As we shall see below, there are individuals who act as if they do not accept the principle of diminishing marginal utility of money: for them the marginal utility increases, at least for some values of x. It is perfectly possible for people to act in this way and yet be coherent. My personal opinion is that people who appear to violate the principle on some occasions are incoherent and would change some of their actions if they realized this, but certainly I would not insist that they accept the principle, at least in the same way that I would insist they should accept the basic axioms or else give me reason why they should not. The same holds for the other properties of $u(x)$ so far examined, for example its increasing with x. They do not follow from the initial premises. They could, of course, be added as additional premises but they would be premises which, for myself, do not have the conviction of those already defended. There is therefore a powerful distinction between the properties of utility described in the present chapter and the laws of probability derived in Chapter 3. The latter are indisputable, the former are merely reasonable. If you violate the one you are wrong: disagree with the other and you are merely unusual. In the language of section 1.4, the laws of probability and the existence of utility are part of the normative approach; whereas the properties of utility now being discussed are prescriptive. You do not have to take the prescription.

5.6 TWO EXAMPLES OF UTILITY FUNCTIONS FOR MONEY

We now have several reasonable properties for $u(x)$: increasing from $u(0) = 0$ to an upper limit of 1 as x increases, and such that $u(x+a) - u(x)$ decreases with x for every positive a. Such a function is tabulated in Table A.I, at the end of the book, and graphed in Figure A.I. A second such function is similarly displayed in Table A.II and Figure A.II. Detailed explanations of these are given alongside them. The reader may wish to interrupt the reading of this chapter in order to study them, since they will be referred to in many of the subsequent examples. We invent two individuals called for the moment I and II (names will be provided in a moment) with these two functions as their respective utilities. Although the two utilities look somewhat similar we shall later see that they differ in at least one important respect, and under similar circumstances the individuals I and II might behave differently owing to their possessing different utilities.

It is easy to see that the two utility functions have all the properties so far suggested as appropriate. They certainly increase from $u(0) = 0$ to an upper limit of 1. It is almost as easy to see that they have diminishing marginal utility. A reference to Figure 5.1 will make it clear. The points A and B are at values of x and $x + a$ dollars, respectively, so that the distance AB represents a dollars. The distance CD also represents a dollars but C corresponds to a much larger sum than does A (or even B). In changing from C to D there is the same marginal increase in money (namely a) as in changing from A to B, but the initial assets are larger. The marginal increase in utility corresponding to AB is PQ: that corresponding to CD is equal to RS. Clearly RS is less than PQ,

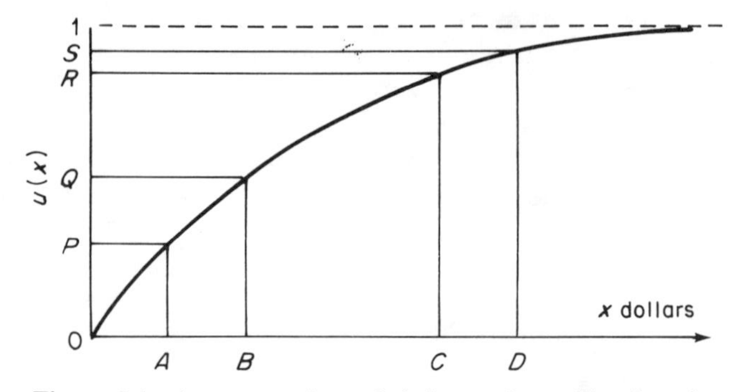

Figure 5.1. A concave, bounded, increasing utility function

that is, the marginal increase in utility at the larger capital sum is smaller than that at the lesser capital; or the marginal utility decreases with capital. An alternative way of expressing the same result is to say that the slope of the graph decreases as x increases—imagine walking up a hill shaped like the graph: you would find it harder at the beginning (small x) than on the almost level plateau at large x. (Those familiar with the necessary mathematical language will appreciate that the second derivative of $u(x)$ is negative.) A function having these properties is said to be *concave*. Our two utility functions are increasing, bounded by the values 0 and 1, and are concave.

From Figure 5.1 it will also be appreciated how $u(x)$ manages always to increase with x and yet never exceeds 1. The marginal utility is always positive and yet gets less and less as the assets increase. Indeed, it can be made arbitrarily small by supposing the assets large enough. To see this, imagine C moved to the right, together with D, so that CD is fixed; then RS diminishes without limit. In other words, 100 dollars is chicken feed for a rich individual.

In using the utility functions given at the end of the book it is often convenient to change the scale of the assets, x, thereby obtaining another function. For example, as it stands, function I attaches a utility of 0.632 to 100 dollars. This is a high value for most people, and it might be more realistic to multiply the horizontal scale by 10, thereby assigning the same value of 0.632 to 1000 dollars and reducing the utility of 100 dollars to 0.095, the value originally found against $x = 10$. These two utility functions are distinct and we can use whichever is appropriate: the tabulation of one effectively provides, by simple changes of scale, a lot of others. Experience shows that many people associate a utility of around 0.5 with their current assets. A function having this value can be obtained as follows. At the moment function I has a utility of one half at $x = 69.4$ dollars, or in round figures, 70 dollars. If someone has a capital of 7000 dollars he might find I an acceptable utility function on multiplying the horizontal scale by a hundred, when, for example, a doubling of capital to 14 000 would give a utility of 0.753, an increase of about 50%. Generally, with a capital of C you might find I an acceptable utility for you if the scale is multiplied by $C/70$. (We shall see below that II, with factor $C/36$, is more likely to be suitable.) Remember that there is nothing mandatory in the use of these functions: they are merely suggestions, and the reader is encouraged to invent and use his own.

5.7 THE INVESTMENT EXAMPLE

To illustrate the use of these utility functions, consider the investment example of Table 5.1. To define the problem completely it is necessary to specify C and the probability of the stock appreciating, $p(\theta_1)$. Let the latter be denoted by p. We study the dependence of the solution on C, p, and whichever of the two utility functions is employed. In all cases the scale of x will be supposed multiplied by 10 for convenience. Suppose first, that $C = 100$, the lowest possible value. (Even here the invested sum, 500 dollars, will have had to have been borrowed. It is presumed that the associated interest charges have been taken into account.) The relevant utilities can be read off the graphs or taken from the tables and are as folows:

	Decision-maker	
Dollars	I	II
200	0.181	0.364
100	0.095	0.221
0	0.000	0.000

Thus, for the first decision-maker, at 200 dollars we enter Table A.I at $x = 20$ (a scaling by 10) and read 0.181. For him, Table 5.1 reads

	θ_1	θ_2
d_1	0.181	0.000
d_2	0.095	0.095

and for the other decision-maker it reads:

	θ_1	θ_2
d_1	0.364	0.000
d_2	0.221	0.221

For I the expected utility of d_1 is $0.181 \times p$, and of d_2, 0.095. These are equal when $p = 0.52$. Hence if the chance of the stock appreciating exceeds this value, he should invest, otherwise not. For the other decision-maker, II, the similar value of p is equal to 0.61. He requires the investment to be safer than the first decision-maker does and it is not until the chance exceeds 61% that the gamble is worth taking.

The low value of C made this an extreme case, so next suppose $C = 1000$

dollars and they are contemplating investing half their fortunes. The relevant decision tables are, for decision-maker I,

	θ_1	θ_2
d_1	0.667	0.593
d_2	0.632	0.632

and for decision-maker II,

	θ_1	θ_2
d_1	0.709	0.676
d_2	0.693	0.693

For I, the expected utility of d_1 is $0.667 \times p + 0.593 \times (1 - p)$, and of d_2 is 0.632. These are equal when $p = 0.52$, the same critical value* as before. A similar calculation for II shows that the two expected utilites are again equal when $p = 0.52$, so that the two decision-makers would behave similarly in this situation, both preferring the investment only if its chance of success exceeds 52%. As a result of his increase in total assets the second decision-maker is prepared to accept the investment at a lower chance than before. Thus if $p = 0.57$, he would not have accepted it with a capital of 100 dollars, but will with one of 1000 dollars. The first decision-maker is not so affected by the change in capital. We leave the reader to explore what happens when C increases still further, say to 2000 dollars.

5.8 UTILITY AND THE AVOIDANCE OF INCOHERENCE

The example concerning four bets, introduced earlier in this chapter, can now be discussed using a utility function. Consider decision-maker II with assets of 50 dollars. He can win or lose 10 or 20 dollars so the relevant utilities are:

Dollars	30	40	50	60	70
Utilities	0.458	0.523	0.570	0.605	0.633

The probabilities of winning that the reader was asked to determine can be found by equating the expected utilities of the bets with the utility of the *status quo*, namely 50 dollars. Thus p_1, for bet I to win or lose 10 dollars satisfies

$$0.523(1 - p_1) + 0.605 p_1 = 0.570$$

* The reader who takes the trouble to perform the calculations described may sometimes find discrepancies of 1 or even 2 in the last place between his results and those in the text. These are 'rounding-errors' and no significance should be attached to them. Thus here he may find $p = 0.53$.

giving $p_1 = 0.57$. The other values, similarly obtained, are $p_2 = 0.43$, $p_3 = 0.76$, and $p_4 = 0.64$. These values do not differ much from those of the incoherent decision-maker cited above—indeed, except for p_2, they agree to the first place of decimals—but the change is adequate to avoid incoherence. To see this it is only necessary to note that each bet now has expected utility 0.570 and therefore each mixture has the same expected utility, unlike the two mixtures A and B considered above.

An alternative method of establishing the incoherence is to demonstrate the lack of a utility function corresponding to the quoted probabilities, $p_1 = 0.6$, $p_2 = 0.5$, $p_3 = 0.8$, and $p_4 = 0.6$. We saw, at the end of the last chapter, that a utility function may have its origin and scale changed without affecting the decisions, so let us suppose that this person making the probability assessments has utilities of 0 and 1 respectively for the worst outcome, a loss* of 20 dollars, and the best, a gain of 20. Then the assertion that $p_4 = 0.6$ for bet IV means that the utility of the *status quo* equals

$$0 \times 0.4 + 1 \times 0.6 = 0.6$$

Bet III, with $p_3 = 0.8$, has expected utility

$$0 \times 0.2 + u \times 0.8$$

where u is the utility of a win of 10 dollars, which again must equal that of the *status quo*, so $u = 0.75$. Bet II, with $p_2 = 0.5$ has expected utility

$$v \times 0.5 + 1 \times 0.5$$

where v is the utility of a loss of 10 dollars. Again, this must equal 0.6 and $v = 0.2$. Finally the first bet has $p_1 = 0.6$ and an expected utility of

$$v \times 0.4 + u \times 0.6$$

which, on inserting the values of u and v just found, gives 0.53 and not 0.6 as it should. Hence there is no utility function which explains the decision-maker's choices and incoherence results. This demonstrates the first point about the example, namely the need for a utility function in reaching decisions involving money. The general approach is in section 9.14.

In the remainder of this chapter we investigate some of the results which follow from having a bounded, increasing, concave function for the utility of money, illustrating the ideas by means of the two utility functions provided at the end of the book. As a by-product of this investigation we shall show that several of the criticisms that have been levelled against utility, and against selecting an act on the basis of expected utility, are not valid.

5.9 RISK-AVERSION

We show that a decision-maker with a concave utility function is, in a sense

* Strictly, we may not talk about the utility of a loss. The phrase used is a convenient abbreviation for 'utility of 0 for the outcome $C - 20$, where C is his current assets'.

now to be described, averse to taking a risk and does not like some apparently favourable gambles.

Consider a decision-maker with the utility function sketched in Figure 5.2 and total assets indicated by the point A in that figure. The utility corresponding to A is indicated by the point labelled P. Now suppose he considers a gamble in which he is equally likely to win or lose a certain sum, for example, the simple investment situation studied earlier in this chapter but with the added condition that the stock be equally likely to prosper or fall; in the former notation, $p = 1/2$. If he wins the gamble his assets increase to C (in the figure), if he loses they fall to B with $BA = AC$ and with $p = 1/2$ the two possibilities are equally likely. Assessed in monetary terms the gamble is fair, in the sense explained in the previous chapter that the expected gain in *dollars* is zero. With $p = 1/2$, this is because A is the mid-point of BC. The situation is different if the gamble is judged in *utiles*. Referring to the figure, the utilities corresponding to A, B, and C are denoted by P, Q, and R. P is above S, the mid-point of QR, because of the property of diminishing marginal utility, PR corresponding to AC being less than QP corresponding to BA. As S is midway between Q and R, representing the two utility outcomes of the gamble, S represents the expected *utility* of the gamble and we see this is less than the utility, at P, of refusing it. Hence, in utiles, the gamble is subfair, and will be refused by the decision-maker. The next two paragraphs present two other ways of looking at the same phenomenon and may be omitted by someone who is convinced by the above argument.

The concavity of $u(x)$ may alternatively be expressed by the fact that the slope of $u(x)$ decreases with x. Referring again to Figure 5.2, where L, M, and N are the points on the utility curve corresponding to monetary sums A, B, and C and utilities P, Q, and R, respectively, this means that ML is steeper than LN. If the straight line, or chord, MN cuts AL in K, so that K is the midpoint of MN, this means that L is above K: hence P is above S and the same argument works.* For a concave curve, a chord MN always lies below the curve.

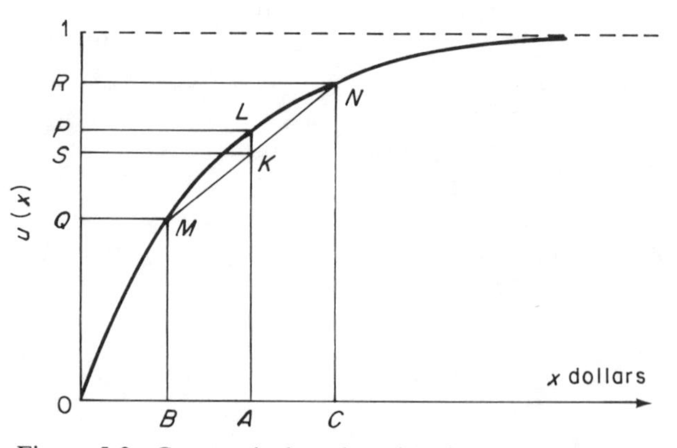

Figure 5.2. Geometrical explanation for risk-aversion

The same result may be obtained without reference to a figure. Let the assets be x and let the gamble have equal chances of winning or losing a, so that, if accepted, it will result in assets $x+a$ or $x-a$. Then because of diminishing marginal utility

$$u(x+a)-u(x)<u(x)-u(x-a)$$

Now add $u(x)+u(x-a)$ to both sides and obtain

$$u(x+a)+u(x-a)<2u(x)$$

and finally multiply by 1/2 to give

$$u(x+a)/2+u(x-a)/2<u(x)$$

The left-hand side is the expected utility of the gamble and the right-hand side the utility of not gambling, so that again we have the result that on the basis of utility it is better not to engage in a monetarily fair gamble.

The result may be generalized to a gamble which has probability p of winning a and complementary probability $1-p$ of losing b, provided that it is monetarily fair; that is, provided

$$a\times p-b\times(1-p)=0 \tag{5.1}$$

(The above was the special case $p=1/2$, $a=b$.) The argument is similar though A will not be the mid-point of BC but will, because of equation (5.1), divide BC in the ratio p to $1-p$. We leave the details to the interested reader: the essential point is that K will be below L and S below P. The result may be further generalized to a bet with several possible outcomes, and not just two. The procedure here is similar to that adopted in section 4.3 when the n possible consequences of decision d_i were reduced to just two, c and C. The several outcomes can be reduced to two without destroying the monetary fairness, and we are back at the original situation. Again details are left to the interested reader.

The upshot of all this is that a decision-maker with concave utility for money will refuse a monetarily fair bet. He is said to be *risk-averse*. In using this last term remember that it refers to monetarily fair bets and he is not averse to a risk with sufficiently high monetary expectation.

5.10 PROBABILITY PREMIUM

We have already carried out some simple calculations to illustrate the point. With the investment that could gain or lose 100 dollars, we saw in section 5.7 that in every numerical case it would not have been made had the chances been equal. In one case the chance would have had to increase to 61% before the investment became worthwhile. In general, let p be the probability that makes

---

* Since S is the mid-point of QR and A the mid-point of BC, K is the mid-point of MN and SK is horizontal.

a bet on two outcomes monetarily fair; and let P be the probability that makes the same bet fair on a utility basis, so that the expected gain in utiles is zero. Then we have shown that P exceeds p. The difference between them, $P - p$, is called the *probability premium* of the gamble. It describes the increase in probability needed to change a fair bet (from the monetary viewpoint) into an acceptable bet (from utility considerations). In the examples considered the probability premium varied between 2% and 11%.

The effect of risk-aversion becomes more noticeable the larger the sums involved are relative to the initial assets. Suppose decision-maker II has total assets of 500 dollars and contemplates a bet which will either lose him everything or win him 500 dollars, so doubling his assets. Then this is monetarily fair if the chance of a win is 1/2. The utilities of 500 and 1000 dollars are, respectively, 0.570 and 0.693 (from Table A.II with a factor of 10) and that of zero is zero. Hence the expected utility of the bet is $1/2 \times 0.693 = 0.347$, whereas if he does not gamble the utility is 0.570. It is obviously better to refuse the bet. For the bet to be worth considering its expected utility must be at least that of not betting, that is, $0.693 \times P = 0.570$, giving $P = 0.82$. He would need to be 82% certain of winning before the bet could be entertained. The probability premium is $0.82 - 0.50 = 0.32$, or 32%. The conclusion here is in reasonable agreement with common sense because this gamble has so much at stake that he would naturally want a high chance of winning before accepting it. Our point is that this natural desire is explained by maximising expected utility with a concave utility function.

The phenomenon of risk-aversion is a result of the assumption of diminishing marginal utility, most clearly expressed in the *concave* form of the graphs in which the slope decreases with the assets. If the curvature of the graph is in the opposite sense, so that the slope *increases* (the curve is said to be *convex*) then the contrary phenomenon exists and the decision-maker will accept risky gambles.

Consider Figure 5.3 in which a utility function is sketched having increasing steepness for small x and only reverting to the familiar form for larger x.

Then in the lower part of the curve a chord MN lies *above* the utility curve, contrary to the case in Figure 5.2, where it lay below it. The effect of this is to reverse the argument given in connection with the earlier figure, consequently the expected utility of a monetarily fair bet exceeds the utility of not gambling and the probability premium is negative. In this part of the curve the decision-maker has increasing (not diminishing) marginal utility. He is said to be *risk-prone*, rather than risk-averse.

It has been suggested that many people have a utility function of the general shape illustrated in Figure 5.3, convex for small assets and concave for larger ones. Personally I doubt this because it would imply a reversal of the decision discussed above concerning a gamble which had equal chances of doubling one's assets or losing everything. If all the assets referred to the convex part of the curve such a gamble would be accepted, whereas I do not think most people would be so wild. Admittedly the point is not easily resolved if only

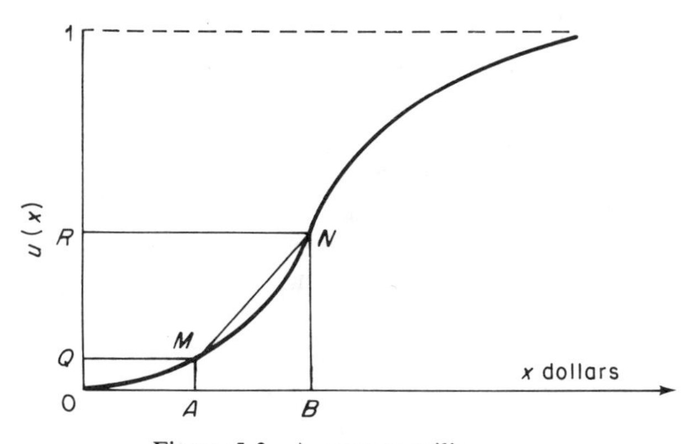

Figure 5.3. A convex utility curve

because it is difficult to see exactly what is meant by zero assets and therefore zero utility. Fortunately few of us ever have occasion to explore the extreme left of our utility curve. (Remember *monetary* consequences are under discussion: death is not being considered.)

5.11 TYPES OF RISK-AVERSION

Although the curvature of the utility function obviously affects the risk-aversion, it is not quite obvious how the amount of curvature is related to the amount of risk-aversion. To illustrate the relationship the two utility functions, I and II, have been provided. Decision-maker I has, in a sense now to be explained, a constant aversion to risk; whereas II has an aversion which decreases as his assets increase. This is despite the fact that the curve in Figure A.I has more curvature for small x than for large x. The calculations already performed will illustrate the difference.

In considering the investment situation in section 5.7 we saw that I required a chance of 52% before investing, whether his assets were 100 dollars or 1000 dollars. If the reader did the calculation suggested, he will have obtained the same value at 2000 dollars. The probability premium is therefore fixed at 2%. On the other hand, for II the premium dropped from 11% at 100 dollars to 2% at 1000; for 2000 it has dropped almost to zero.*

For I, the premium, for a given bet, does not depend on the assets: nevertheless, for given assets, it does change with the bet. For example, at 1000 dollars assets, we saw that the premium for a bet to win or lose 100 dollars is 2%, whereas to win or lose 1000 dollars it is easy to calculate the value at 23%. (Remember that we were scaling the money by 10 before using Table A.I: the utility of 1000 is obtained by entering the table at $x = 100$, giving 0.632.)

Decision-maker I will be referred to as the constantly risk-averse decision-maker, meaning that the probability premium for a gamble does not vary with

* It is not possible to make very precise calculations at $C = 2000$ due to the limits of accuracy of the tables.

his assets. II will be called a decision-maker with decreasing risk-aversion, meaning that the probability premium for a gamble decreases as the assets increase. Most of us are decreasingly risk-averse, as can be seen from commonsense consideration of the investment example, where a person would typically react differently to the investment if his capital, C, changed and a rich person would take greater risks than a poor one. A utility function like that of Table A.II is therefore more realistic than that of Table A.I. Of course, there are several utility functions which yield decreasing risk-aversion and that in Table A.II is merely an example. It does so happen, however, that the only constantly risk-averse utility is that given in Table A.I, apart from a possible change in scale. The following paragraph explains this point in more detail but may be omitted by the reader unfamiliar with the mathematical language.

The probability premium for a bet to win or lose h is given by $P - 1/2$ where P satisfies

$$u(x+h)P + u(x-h)(1-P) = u(x)$$

Hence

$$P = \{u(x) - u(x-h)\}/\{u(x+h) - u(x-h)\}$$

For $P - 1/2$ not to depend on x we must have

$$-(P - 1/2) = \frac{u(x+h) - 2u(x) + u(x-h)}{2\{u(x+h) - u(x-h)\}}$$

constant. In particular, the limit of this as h tends to zero, namely $u''(x)/u'(x)$ must be constant, where the primes denote differentiation. The solution to this differential equation with $u(0) = 0$ and upper bound one is

$$u(x) = 1 - e^{-cx}$$

for some positive c. c is clearly a scale factor on x and the essential uniqueness of the constantly risk-averse utility is established. Notice that

$$\log_e\{1 - u(x)\} = -cx$$

so the disutility, $1 - u(x)$, has its logarithm linear in x. There is a much wider class of decreasingly risk-averse utilitites: for example, any function of the form

$$1 - we^{-ax} - (1-w)e^{-bx}$$

with $a > 0$, $b > 0$, $0 < w < 1$ has the property, and there are many others. Table A.II gives the function

$$1 - e^{-x/200}/2 - e^{-x/20}/2$$

5.12 INSURANCE

We now leave the mathematics and return to the main discussion. So far we have been discussing whether or not to accept a gamble; whether to invest the money or keep it in the bank; whether a company should try to enter a new field or stay in its own; whether a risky situation is preferred to a non-risky one. The conclusion we have come to is that, with diminishing marginal utility of money, a fair risk from a monetary point of view is unacceptable and a probability premium is required to make it worth considering. The amount of this premium depends on the risk, and usually on the assets, decreasing as the assets increase. Let us now consider the position where we have a risky situation and are considering disposing of it. The usual way of doing this is through insurance. If we own a house, that house may be destroyed by fire or other calamity, but we can take out an insurance which will compensate us in the event of the calamity at the cost of an annual premium paid to the insurance society. From a decision table viewpoint the situation is described in Table 5.2. (The table is again simplified; partial loss of the house is another possibility, but we do not wish to complicate the analysis.) If the total assets are C, the value of the house h, and the premium m, the monetary consequences are as in Table 5.3 (cf. Table 2.4). The inconvenience of temporarily losing the house is supposed compensated for, so that the insurance leaves the decision-maker in the same position had the calamity not occurred. Here m is much less than h, and p, the probability of θ_1, the chance of a calamity, is small.

Now Table 5.1 (for the investment problem) and Table 5.3 are very similar. In both cases there is one decision whose outcome is certain (d_2 in Table 5.1, d_1 in Table 5.3) and another which is risky. The difference between the two tables is that in the former inactivity results in choosing the certain situation (leave the money in the bank) whereas in the latter it results in the risky situation (possible loss of an uninsured house), but as a problem of choice between two decisions this difference is immaterial. Consequently many of our remarks about the earlier situation apply to the present one. In particular, a risky situation is avoided by a risk-averse decision-maker: therefore he will tend to opt for insurance in order to remove the risk.

Interest in the insurance situation centres on the premium quoted by the

Table 5.2

	θ_1: Calamity	θ_2: No calamity
d_1: Insurance	Inconvenience, but otherwise *status quo*	Loss of premium
d_2: No insurance	Loss of house	*Status quo*

Table 5.3

	θ_1	θ_2
d_1: Insurance	$C-m$	$C-m$
d_2: No insurance	$C-h$	C

company and only if this is low enough will the insurance be undertaken. We therefore consider a numerical illustration of Table 5.3 and see how the reasonable premium might be found. Suppose $C = 5000$ dollars and that the calamity represents total loss: that is, $h = 5000$ and $C - h = 0$. Suppose we take the decreasingly risk-averse decision-maker and scale Table A.II by a factor of 100 (not 10, as before). The utility of 5000 dollars is then found with $x = 50$, namely 0.570. If p is the chance of the calamity, the expected utility without insurance (d_2) is $0.570 \times (1 - p)$: with insurance it is $u(C - m)$. Consequently an acceptable premium, m, satisfies

$$u(C - m) = 0.570 \times (1 - p)$$

with $C = 5000$. Any smaller value of m will, of course, be acceptable. If $p = 0.01$, so that there is just one chance in 100 of calamity, we have

$$u(5000 - m) = 0.570 \times 0.99 = 0.564$$

Using Figure A.II or Table A.II in the reverse direction and finding the value of x with a utility of 0.564, gives about 48.6, or $5000 - m = 4860$ and $m = 140$ dollars. Hence the decision-maker would be prepared to pay up to 140 dollars premium against the total loss of his assets.

This premium of 140 dollars is much more than the actuarial value of the loss based on expected monetary values. 5000 dollars with a chance of 0.01 of loss gives an expected monetary loss of only 50 dollars. Hence the premium could be about three times the actuarial loss. This suggests that insurance would not only be desirable at 140 dollars, but that a society could be found to offer this premium. The total assets, C' say, of an insurance society will be large in comparison with the assets C of the insured and if the society accepts insurance it will be involved with possible consequences $C' + m$, C', $C' + m - h$ and, since C' is much larger than h or m, these three consequences will be similar. Hence on the society's utility curve the three possible consequences will correspond to points that are very close together. For example, 5M (M for million) might be a reasonable value for C' when the other values would be 5000140 and 4995140. A scale factor of 100000 might enable Table A.II to be used with $x = 50.0014$, 50 and 49.9514. Over this small range the curvature of $u(x)$ is slight and $u(x)$ is effectively linear in x. We saw that a linear transformation of utility had no effect on decision-making so we can, over this range, take utility to be the same as money. Hence the company will view the situation actuarially, and a premium of over 50 dollars will be acceptable to them. In addition, they have administrative costs, but even if these equal the actuarial costs a premium of 100 dollars will cover both and, being below 140, be acceptable to the individual. Hence the concavity of the risk function explains why insurance works. The individual is operating at a curved part of his utility graph, whereas the society's graph has no appreciable curvature over the range considered and it can equate utility with money. It follows as a corollary that if the curvature for the individual is very small, he will not find it possible to obtain attractive premiums. It does not pay to insure against small losses, only against large ones.

The difference of 90 dollars between (1) the expected monetary loss associated with the calamity and (2) the amount the individual would be prepared to pay to dispose of the gamble, is called the *risk premium*. Like the probability premium defined earlier in this chapter it measures the difference between the monetary and utility assessments of the gamble. The probability premium expresses it by describing by how much the probability would have to change to make the gamble acceptable: the risk premium achieves the same end by stating what variation in the rewards (or losses) is necessary at fixed probability.

5.13 EXPECTED UTILITY AS THE SOLE CRITERION

An objection that is often levelled against the use of expected utility in reaching decisions is that it is unreasonable to judge an action d_i on the basis of just one number: its expected utility, $\bar{u}(d_i)$. It is argued that there are many facts to be considered in connection with each action and it is not sensible to expect to be able to describe them adequately by a single value. The following two examples will illustrate the point.

A scientist in charge of an industrial research laboratory has the task of selecting the projects to be investigated under his guidance. Some projects are fairly standard and he can say with a fair degree of confidence that, if attempted, they will most likely be successful and produce a useful, but modest, return. An improvement in an existing product might come into this category. On the other hand, there are some projects which are unpredictable and hazardous, but if they come off make a fortune for the firm. A drug successful against a killing disease might be an example. The scientist will often argue that he must take into account not only what can be expected to happen with a project, but must also consider the greater hazards associated with some projects. He might argue that the two types of project just quoted have the same expectations, but that the second might land the firm in financial difficulties if it failed, and that this great variability in the possible outcomes makes it less desirable than the former. It is better to be safe than sorry.

A second example arises in advising on the selection of a portfolio of stocks. Some stocks, gilt-edged for example, are fairly sure to give a modest yield whilst others are risky and might give a handsome return, or nothing. Surely, it is argued, the variability of the latter needs to be considered in addition to its expected yield.

The counter-argument is that the expected utility does take the riskiness of the action into consideration and that the objection only arises because of a confusion between expected utility and expectations of other quantities, for example, money. It must be remembered that utility is not just a number describing the attractiveness of a consequence but is a number measured on a probability scale and obeys the laws of probability. For example, the square of utility would be a legitimate measure to associate with a consequence, in the sense that the more desirable the consequence, the larger the measure, but the square would not obey the laws of probability and therefore we could not justify using its expectation. It is this unique probability interpretation of utility that makes its expectation so important. When, in addition, the utility function is concave, the increasing riskiness of a gamble is reflected in a corresponding diminution in expected utility, without a corresponding fall in expected financial reward, in the way we have already described by risk-aversion. The previous numerical examples will be sufficient to illustrate the point.

Consider the decision-maker of constant risk-aversion (and a scale factor of 10) with assets of 100 dollars contemplating the investment which is equally likely to win or lose him 100 dollars. The expected utility of the investment is

$$1/2 \times 0.667 + 1/2 \times 0.593 = 0.630$$

whereas, left in the bank, the utility is 0.632. The riskiness of the venture has resulted in a small loss of expected utility and the safer course is preferred. Hence the effect of riskiness is included when employing expected utilities.

Now take a slight variant of this problem in which the retention of the money in the bank is certain to result in a loss of 6 dollars, all the other factors remaining the same. The utility of $1000 - 6 = 994$ dollars is 0.630, the same as the risky investment. The 6 dollars is another example of the risk premium recently referred to in connection with insurance and describes by how much the safe course of action has to be reduced to bring it to the level of the risky venture. More risky ventures have lower expected utilities and higher risk premiums. Thus, still with assets of 1000 dollars, the 'fair' bet to win or lose 1000 dollars has expected utility

$$1/2 \times 0 + 1/2 \times 0.865 = 0.432$$

much less than 0.632 obtained by leaving it in a bank. Using Table A.I (or Figure A.I) in reverse we see that 566 dollars is the sum having a utility of 0.432. Hence the risk premium is $1000 - 566 = 434$ dollars.

There is another reason for thinking that no more than one number is needed to select the best decision, namely that in general, we cannot maximize more than one thing at a time. Suppose, for example, that instead of one number, each decision, d_i, has associated with it two numbers, u_i and v_i, and that the larger u_i or v_i the better was the decision. Thus u_i might be the expected utility, formerly denoted $\bar{u}(d_i)$, and v_i might be some measure of the riskiness, the larger v's being associated with the smaller risks. Then how is the best decision to be selected, granted that one wants to make u_i and v_i as large as possible? Figure 5.4 illustrates the situation with just three decisions, each point corresponding to a decision. It is clear that d_3 is no good because d_1 and d_2 have values of both u and v larger than those of d_3 (u_1 and u_2 exceed u_3: v_1 and v_2 exceed v_3). Thus d_3 can be rejected and the choice lies between d_1 and d_2. The selection here is genuinely difficult because d_1 gains over d_2 in respect of u(u_1 exceeds u_2) but d_2 is better than d_1 on the basis of v (v_2 exceeds v_1). In

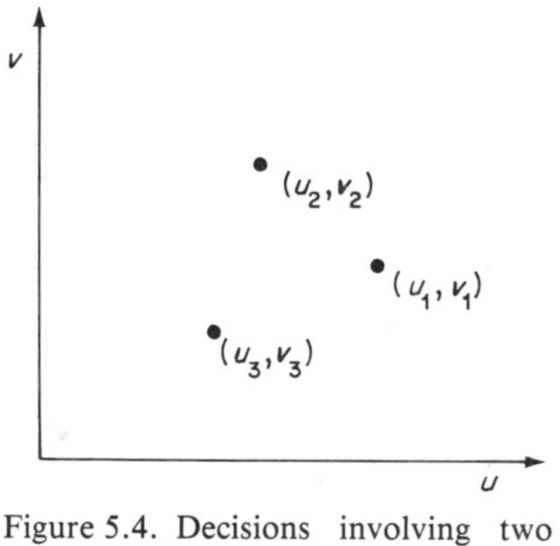

Figure 5.4. Decisions involving two considerations, u and v

deciding between them, the benefits of u increasing have to be balanced against the loss in v when changing from d_2 to d_1 and until some such balance is struck, no solution is available. A decrease in riskiness has to be balanced against a loss of expectation. Expected utility does just this, and its maximization solves the decision problem.

5.14 PORTFOLIO MANAGEMENT

As an example of a situation where it is necessary to consider both risky and relatively secure ventures, we cited the problem of an investor choosing a portfolio of stocks, and we claimed that the expected utility of a stock is an adequate description of the stock for decision purposes. In compiling a portfolio advice is often given to diversify one's holdings by putting some money into securities and some into more speculative material. How far is this advice sound? It is sometimes argued that the principle of maximizing expected utility goes against it, because each stock has its expected utility and therefore, applying the principle, the best thing to do is to put all one's money into the stock of highest expected utility. We now show that this last argument is facile and that a more careful investigation shows that the maximizer of expected utility, the coherent decision-maker, will typically act on the advice and diversify his investments. To establish that the advice is sound in general requires some mathematics, so we will content ourselves with illustrating the phenomenon in a simple situation.

The example involves our decision-maker with constant risk-aversion (Table A.I) contemplating investing in the stock described at the beginning of the chapter, or leaving his money in the bank. We suppose he has a capital of 1000 dollars and his utility is as in Table A.I with a scale factor of 10. The stock is such that every 500 dollars invested will increase to 600 with probability p, or decrease to 400 with probability $1 - p$; smaller or larger amounts will increase or decrease in proportion. Money left in the bank will keep its value.

We suppose $p = 0.55$. In our earlier calculations we saw that with this value (greater than the critical value of 0.52 needed to overcome the risk-aversion) it was better to invest 500 dollars than to leave it in the bank.

Now we ask the new question: what is the optimum amount to invest in the stock? We have seen 500 dollars is better than none, but perhaps it might be best to risk all 1000 dollars. The following table lists the possible outcomes, in dollars, for five typical investments between 0 and 1000.

Amount invested in stock	Stock appreciates	Stock depreciates
0	1000	1000
250	1050	950
500	1100	900
750	1150	850
1000	1200	800

One way of looking at this is as a decision table with five decisions and two uncertain events. (The real decision table has 1001 decisions if investments can be made in dollar units.) From this table we can form a decision table with probabilities and utilities (from Table A.I) and hence expected utilities. This is shown in Table 5.4. The final column shows that the investment of highest expected utility amongst those considered is 500 dollars. More refined calculations, or use of the differential calculus, show that the true maximum is very close to this value. In other words, the optimum procedure for the coherent decision-maker is to invest half his capital in the stock, leaving the other half in the bank; that is, to diversify his holdings between the secure and the risky. (We leave the reader to verify that it is not until the probability of the stock appreciating increases to about 0.60 that it is worth investing all his capital in it.) The advice is therefore sound, though the increase in utility of 0.002 between the best and worst decisions is small, corresponding to a monetary change of around 5 dollars. With other examples larger differences can arise. The facile counter-argument given above fails to work, again because it ignores the diminishing marginal utility reflected in the curvature of the utility function.

Table 5.4

Amount invested in stock	Stock appreciates	Stock depreciates	Expected utility
0	0.632	0.632	0.632
250	0.650	0.613	0.633
500	0.667	0.593	0.634
750	0.683	0.573	0.633
1000	0.699	0.551	0.632
Probability	0.55	0.45	

5.15 REPEATED DECISION PROBLEMS

It has already been explained in section 2.4 that the approach to decision-making described here is relevant to a single occasion on which a decision is to be taken, and does not depend for its validity on the real or conceptual repetition of the same problem, as do some methods. The uniqueness of the decision-making occasion is emphasized in the probability considerations, which invite comparison of the uncertain event with the single drawing of a ball from an urn. Nevertheless there are situations in which essentially the same decision problem occurs repeatedly. A simple example is provided by the manufacturer, quoted in section 4.1, faced with the problem of whether or not to inspect material before its despatch to the customer. Presumably he is producing the material in large quantities and the same problem will present itself with each roll so that there is genuine repetition of the same decision problem.

Anything like a reasonably complete discussion of this topic is impossible because many of the problems it raises are difficult. We therefore content ourselves with illustrating one type of repetition: first, in order to show how something rather interesting and perhaps surprising can occur; and second, to dispose of another objection to the principle of maximization of expected utility. The objection says that this principle, if applied whenever a particular situation arises, will always lead to the same choice of act and therefore a stimulus will always produce the same response. It is held that this is absurd as a general rule of action. We agree with the absurdity of the conclusion but demonstrate that it does not follow from the maximization of expected utility.

The illustration is simple and the reader is warned to be cautious in extending the result to more complicated situations. Suppose a decision-maker contemplates a gamble which will either win for him a given sum with probability p or lose him the same sum with probability $1 - p$: the investment situation just considered provides an example. Then by the arguments already given he can calculate the critical value of p which is just enough to make the gamble attractive and counter his risk-aversion. Now suppose he is offered a second gamble which is just the previous one played over twice: will his attitude towards this gamble be the same as to the single gamble? (If the manufacturer has two items to despatch, will his attitude be the same as for a single item?) It is not hard to show that for a decision-maker with constant risk-aversion the two gambles will be treated alike; if one is accepted, so is the other and the critical values of p will be the same. However, we saw that it was realistic to think of people as having decreasing risk-aversion and here the answers are more interesting.

Specifically, suppose decision-maker II has assets of 50 dollars and that he is offered a gamble with chance $p = 0.678$ of winning 25 dollars and complementary chance 0.322 of losing 25 dollars. The expected utility of the gamble is

$$0.678 \times 0.645 + 0.322 \times 0.415 = 0.571$$

which exceeds 0.570, the utility of refusing the gamble and keeping 50 dollars, and the single gamble should just be accepted. Next consider the double gamble; if he wins on both plays he will win 50 dollars, finishing up with 100, and this has probability p^2, by the product law. (Notice that we are assuming that the two plays are independent, both in the sense that the chance of a win on the second play is unaffected by one on the first, and also in the sense that the decision-maker does not have the option of withdrawing after one play but commits himself to both plays or neither.) Similarly, if he loses on both plays he will lose 50 dollars, and this has probability $(1-p)^2$. The final possibilities are that he wins on the first and loses on the second, with probability $p(1-p)$, or loses on the first and wins on the second, also with probability $(1-p)p$: in either case he finishes up where he started. The expected utility is therefore

$$(0.678)^2 \times 0.693 + 2 \times 0.678 \times 0.322 \times 0.570 = 0.567$$

(the utility of zero being zero) which is less than the utility, 0.570, of refusing the gamble. Consequently although the single play is worth accepting, the double play is not. It is possible to provide examples which go in the opposite direction, that is, where the double play is acceptable but the single play not. Exercise 5.8 at the end of the chapter does just this. Thus we see that decision rules for individual repetitions of the same situation may differ from those appropriate for a single occasion.

5.16 SOME COMMENTS ON MONETARY UTILITY

It has been explained that different persons may have different utility functions and, in particular, one may be more risk-averse than another. Equally, a person's utility function may change with time. Like a probability, to which we have seen it is intimately related, utility depends on the circumstances at the moment of its assessment and may change with those circumstances. In solving a decision problem it is the utilities at the time of choice that are relevant, not those that obtain when the consequence is realized. I might decide I have high utility for a holiday 'away from it all', only to find when I get there that it is not as attractive as I had expected. Little seems to be known about the way utilities change with time, though the manner in which probabilities change is well understood and is the topic of the next chapter.

In analysing some of the situations of the present chapter the reader may have felt himself even more risk-averse than the two people we have imagined: he might, for example, say he would not risk half his assets (500 dollars) on a venture having a 48% chance of losing him 100 dollars. Equally, the same person may find himself less risk-averse in liking to have a gamble on a horse race. This may be explained by changing the utility function for money, but there is another possibility. We have been discussing the utility of money alone and there are many situations in which money is not the only relevant factor. The person considering the possible loss of 100 dollars may feel that the mental anguish to him of the loss, and the possible criticism by his spouse, might well result in lower utility than is suggested by purely financial considerations. The gambler at the racetrack may look upon the sacrifice of the stake as a legitimate price to pay for the added thrill that comes from watching a race in which he has a monetary interest. It is hard to generalize about the utilities of such pains and pleasures, but they should not be forgotten.

Notice that the considerations of this chapter lead to the conclusion that everything has its price. For all consequences have utility, money has utility, and therefore any consequence has, through utility as an intermediary, a monetary equivalent. This is sensible. What is not necessarily true is that the possession of the money will enable the consequence to be reached. 2000 dollars may not enable the good health to be purchased. Rather we say that were we to have the good health it would be equivalent to having an extra 2000 dollars.

Notice the use of the subjunctive 'were' in the last sentence. Its occurrence here with utility is for exactly the same reason that it was used with probability at the end of section 3.11, namely that the situation does not have to be realized, only contemplated. It is not necessary to know the defendant is guilty, we need only consider the possibility of his being guilty. Here we do not need to have the consequences, money or good health, only contemplate them. Indeed, throughout the discussion of the utility of money, the values obviously only require contemplation. For example, we have only to think about our risk-aversion were we to have 100 dollars.

Exercises

5.1. The decision-maker with decreasing risk-aversion has assets of 30 dollars and has a decision problem with the following structure:

	θ_1	θ_2
d_1	-10	$+5$
d_2	$+15$	-5
pr	0.3	0.7

The entries in the table represent gains or losses in dollars: thus with d_1 and θ_1 he will finish up with assets of 20 dollars. Advise him on the choice of decision. Would your advice remain the same if his assets were 200 dollars?

5.2. The same decision-maker as in Exercise 5.1 has assets of 20 dollars and contemplates a gamble which may win him 20 or lose him 10 dollars. It is therefore actuarily fair if the chance of winning is 1/3. Determine his probability premium. Do the same for the constantly risk-averse decision-maker. Find the probability premium for the first person when his assets are 200 dollars.

5.3. Perform the exercise suggested in section 5.7: that is, for each decision-maker find the probability premium with assets of 2000 dollars and a chance to win or lose 100. (Remember both had a scale factor of 10.)

5.4. The decision-maker with decreasing risk-aversion has assets of 50 dollars and realizes that 25 dollars of it is in a risky situation which has 1 chance in 10 of collapsing and losing him his money. What is a reasonable premium for him to pay for insurance against the loss? What would the premium be if his assets were 100 dollars?

5.5. A decision-maker has the following utilities for money:

Money	0	20	40	60	80	100	120	140
Utility	0	0.04	0.13	0.27	0.50	0.73	0.87	0.96

By sketching a graph or otherwise show that he is risk-averse for assets above about 80 dollars but not below this amount. Consider a gamble that might win or lose 20 dollars, first when his assets are 40 dollars, and then when they are 100. In each case determine the probability premium.

5.6. An insurance company has assets of 40M dollars and its utility is given by Table A.II with a scale factor of a million (M) dollars. It is asked to insure against an earthquake disaster of 40M dollars and assesses the chance of the earthquake at 0.05. What is a reasonable premium for it to request? (Remember you are looking at the situation from the company's side, not from that of the earthquake-sufferers. Administrative expenses can be ignored.)

The company finds a second company with the same assets and utility function and agrees to share the risk. That is, it insures against a loss of 20M dollars with probability 0.05. What is the premium now? Show that even twice this premium is less than the original premium so that the insurance is improved from everybody's point of view by the companies' sharing the risk.

5.7. The decision-maker with decreasing risk-aversion has assets of 50 dollars. How much would he be prepared to pay for a ticket that entitled him to a 50–50 chance of winning 20 dollars? Suppose he had the ticket already: how much would he be prepared to sell it for? Repeat the question for the constantly risk-averse decision-maker and show that in his case the buying and selling prices are the same.

5.8. A decision-maker has the following utilities for money:

Money	0	10	20	30	40
Utilities	0	0.10	0.17	0.22	0.26

With assets of 20 dollars he contemplates a gamble to win 10 dollars with probability 0.58 or otherwise lose 10 dollars. Show that the gamble should be declined. He then goes on to consider a gamble which consists of two plays of the original gamble. Show that the new gamble should be accepted.

Chapter 6

Bayes' Theorem

'It's these people with fixed ideas who are the danger.'

To Let. Ch. 5.

6.1 INFORMATION AND DECISION-MAKING

A natural reaction of anyone having to make a decision under uncertainty is to try to remove the uncertainty by finding out the true state of affairs. In the language and notation developed in this book, one way of selecting a decision from the set $(d_1, d_2, \ldots d_m)$ when one of the uncertain events $(\theta_1, \theta_2, \ldots \theta_n)$ obtains is to find out which θ_j obtains. For if this is done and the true situation found to be θ_k; then the decision-maker need only consider the consequences $C_{1k}, C_{2k}, \ldots C_{mk}$ and select the one with highest utility: that is, choose that d_i which maximizes $u(C_{ik})$, and obtain utility

$$\max_i u(C_{ik})$$

By removing all uncertainty from the situation, the probabilities and expected values are no longer needed. Furthermore, it is typically true that the decision-maker will be better off than he would have been without the knowledge of the true state of affairs (section 7.4).

This is certainly one way out of some of the difficulties, but unfortunately it is rarely practicable. For example, if θ_j refers to the future and the decision has to be taken now, there is no means of ascertaining the correct event. The investor of the previous chapter could not know for certain whether the stock was going to appreciate or not. Even when it does not refer to the future there may be difficulties in determining θ_k in the context of the decision problem. Thus the manufacturer of section 4.1 trying to decide whether or not to inspect his material before despatching it to a customer knows only that the material is good (θ_1) or bad (θ_2). He cannot determine which except by inspection, which is what he is undecided about, and the removal of uncertainty is not a course open to him. Another consideration that may prevent the uncertainty being removed is simply that of cost, for it may be too expensive to find the truth. An engineer designing a plant may think it useful to build a prototype to resolve some of the uncertainties in the design problem, but the prototype may itself cost so much that it is better to build the plant in a fairly flexible manner and learn from experience what modifications are needed.

So although it is usually a good thing to remove uncertainty, it is often not practical. There is, however, a partial solution. It may not be possible to remove all the uncertainty but sometimes it is feasible to reduce it by obtaining relevant information. Thus the investor may consult his stockbroker, who is presumably more knowledgeable than he. The manufacturer may consult his records and see what sort of proportion of the material has been poor in the past. The engineer, deprived of his prototype, may obtain some information from calculations of the likely behaviour of the system.

The decision-maker has initially probabilities $p(\theta_1)$, $p(\theta_2)$, ... $p(\theta_n)$, expressing the uncertainties, and the effect of additional information will be to revise these probabilities. If he obtains complete information, one of the probabilities will go to 1, namely that corresponding to the true θ_j, and the remainder will go to zero. Partial information will produce a less marked change. If X denotes the information, the revised values will be $p(\theta_1 | X)$, $p(\theta_2 | X)$, ... $p(\theta_n | X)$. We begin this chapter by discussing the following problem: how are the probabilities $p(\theta_j)$, available before the information is obtained, related to the $p(\theta_j | X)$ available after it has been collected? All these values must satisfy the basic rules of probability and be coherent one with another. The coherence is achieved by means of Bayes' theorem, which was proved in section 3.10 and illustrated in section 3.11. (The reader may like to refresh his memory of these two sections before proceeding.)

The theorem says that, for any two events E and F, E not having probability zero,

$$p(F | E) = p(E | F)p(F)/p(E)$$

In the decision-making context replace the event F by θ_j and E by the information X. (The legal example of section 3.11 used information, the preferred word being 'evidence'.) Bayes' theorem then reads

$$p(\theta_j | X) = p(X | \theta_j)p(\theta_j)/p(X) \tag{6.1}$$

immediately relating $p(\theta_j | X)$ and $p(\theta_j)$. These are the probabilities required in solving the decision problem by maximizing expected utility, the former with, and the latter without, the additional information. We say 'additional' here because there already exists background information H omitted from the notation.

6.2 LIKELIHOOD

The role of Bayes' theorem in establishing the connection between $p(\theta_j)$ and $p(\theta_j | X)$ can better be appreciated by using a technical device. The result holds for every θ_j which, remember, are exclusive and exhaustive, so that their probabilities, whether given H or given H and X, add to 1. Suppose, then, we forget $p(X)$ on the right-hand side of the statement of the theorem and merely form the product

$$p(X \mid \theta_j)p(\theta_j)$$

These numbers will not usually add to 1: suppose they add to a value k, say. Then all we have to do is to divide them all by k when they will add to 1. In fact* k is $p(X)$ and on division by k the results will be $p(\theta_j \mid X)$. Hence $p(X)$ is irrelevant; we can form the products and then scale them to add to 1. We write Bayes' theorem in the form

$$p(\theta_j \mid X) \propto p(X \mid \theta_j)p(\theta_j)$$

where the sign $\propto$ is read 'is proportional to'. Now the relationship between $p(\theta_j \mid X)$ and $p(\theta_j)$ is clear: the former is essentially the latter multiplied by $p(X \mid \theta_j)$.

It is convenient to introduce some nomenclature. $p(\theta_j)$ is called the *prior* probability of θ_j: $p(\theta_j \mid X)$ is called the *posterior* probability of θ_j. These are prior and posterior to the information X, which strictly ought to be mentioned explicitly but is usually clear from the context. In doubtful cases we say probability of θ_j prior (or posterior) to X. The omission of X in the wording avoids complexities, as does the omission of H in the notation, but neither should be forgotten.

Both these expressions are probabilities of θ_j and obey the rules of probability described in Chapter 3. The third quantity in the simplified form of Bayes' theorem, $p(X \mid \theta_j)$, is *not* a probability of θ_j (though it is of X) and therefore deserves another name. It is called the *likelihood* of θ_j. Again reference to X is omitted in the wording. Notice how different the likelihood of θ_j is from the posterior probability of θ_j; the arguments θ_j and X have their roles reversed in the two ideas. The example of section 3.10 concerned with male and female death-rates is relevant. Notice that the language here used employs a common mathematical device in that words in the English language which are not very precisely defined are given precise meanings: here probability and likelihood. But the near-synonymity in English is changed to a very marked difference in the mathematical usage. With these terms Bayes' theorem can be stated:

> The posterior probability is proportional to the prior probability multiplied by the likelihood.

So the answer to our question as to how $p(\theta_j)$ and $p(\theta_j \mid X)$ are related is to say the latter is the product of the former with the likelihood, the results being standardized to add to 1.

With only two possibilities, θ_1 and θ_2 ($m = 2$), Bayes' theorem can be written

* This follows also from the extension of the conversation from X to include the θs (section 3.8)

$$p(X) = \sum_{j=1}^{n} p(X \mid \theta_j)p(\theta_j)$$

even more simply, as was done with the legal example in section 3.11, in the odds form

$$0(\theta_1 \mid X) = \frac{p(X \mid \theta_1)}{p(X \mid \theta_2)} 0(\theta_1)$$

since $\theta_2 = \bar{\theta}_1$. The ratio here is called the *likelihood ratio*.

6.3 AN EXAMPLE

This example is designed both to illustrate Bayes' theorem and to show how people have difficulty in assimilating new information coherently so that there is a real distinction between the descriptive and prescriptive views (section 1.4).

Suppose that you have in front of you an urn, such as was used as a standard for probability (section 2.5), that contains a very large number of balls all identical except for colour. You know that either 2/3 or 1/3 of the balls are white and the rest black. In other words, there are either two white balls to every black one (θ_1), or two black balls to every white (θ_2). The possible urns will be referred to, respectively, as the white and black urns, the adjectives corresponding to the compositions of the urns, not to their outward appearances. Next, suppose that you do not know which urn it is that you have but think it is just as likely to be the white as the black. Then $p(\theta_1) = p(\theta_2) = 1/2$, omitting H, which is the rest of the information in this paragraph. Alternatively, the odds for white (or black) are evens.

Suppose that you want to determine which urn it is. The obvious procedure is to tip out the contents onto a table and count the balls: even a glance may suffice to determine which colour predominates. Suppose that this is denied to you, like knowledge of the future, but what you can do is to take out a ball at random (section 2.5) and inspect its colour. This will give you information about the composition. Suppose further you can do this several times: what effect will the knowledge of the colours of the withdrawn balls have on your uncertainties of θ_1 and θ_2?

Specifically, suppose 33 balls are drawn, 20 are found to be white and 13 black. What are the odds in favour of the urn being the white one with predominately white balls? You are asked to consider this problem before proceeding, without considering the technicalities of Bayes' theorem.

6.4 PROBABILITY CALCULATIONS

Having done that, let us use the rules of probability to obtain a coherent answer. What probabilities do we have initially? Let w be the event that the first withdrawn ball is white and b that it is black. Then

$$p(\theta_1) = p(\theta_2) = 1/2, \text{ or } 0(\theta_1) = 1;$$

$$p(w \mid \theta_1) = 2/3, \; p(b \mid \theta_1) = 1/3;$$

and

$$p(w \mid \theta_2) = 1/3, \; p(b \mid \theta_2) = 2/3$$

The first line reflects your opinion of which urn it is: the other lines describe the possible compositions of the urns. By Bayes' theorem in odds form the effect of withdrawing a white ball ($X = w$) is

$$0(\theta_1 \mid w) = \frac{p(w \mid \theta_1)}{p(w \mid \theta_2)} 0(\theta_1) = 2$$

and of a black is $0(\theta_1 \mid b) = 1/2$. That is, the removal of a white ball doubles the odds for the white urn, whereas a black ball halves it. In the language of the previous section, the white ball takes prior odds of 1 to posterior odds of 2, the likelihood ratio being 2.

Now take a second ball. The situation is exactly the same (remember it was assumed that the urn contained a very large number of balls so that the removal of a few will not affect the composition) except that the new odds for the urn being white are 2 if the first withdrawal was white. (The odds *prior* to the second ball being drawn are precisely those *posterior* to the removal of the first ball.) In particular the likelihood ratios are still 2, for a white withdrawal, and 1/2 for the black one. This last statement holds for all subsequent withdrawals so that Bayes' theorem says that each white ball multiplies the odds by 2, each black ball halves them. For example, a black ball followed by a white returns the odds to their initial value of 1.

In our example, 20 balls were white, 13 black. We double the odds 20 times, halve them 13; the overall result being to double them 7 times, equivalent to multiplying by 128. Consequently the answer to the question is that as a result of seeing 20 white and 13 black balls, the odds in favour of the white urn (or against the black, in bookmakers' terminology) are 128–1. This practically means that you are almost certain the urn is white. So here we have an example of coherence; if your uncertainties about the urn initially and about the balls are as stated, you *must* have the final odds of 128–1.

Now compare this value with the one you gave to the question before doing the calculations. Experience shows that the calculated value is usually higher than the assessed one. People commonly give values around 10 to 1. In other words, they under-react to the information: the data, 20 white, 13 black, say more than intuition suggests. I am not suggesting that people always under-react to data. There are familiar occasions where a little anecdote can shift peoples' views dramatically. No: the point is that there are serious discrepancies between description and prescription. This fact makes the coherent view most important because our present incoherent, inefficient ways can be improved by using it.

6.5 SUFFICIENCY

Notice that the calculation of the posterior odds does not use all the information of 20 white and 13 black balls but merely the fact that there were 7 more white than black: only the difference is relevant. This is because each white ball multiplies the odds by 2, each black divides by 2. Consequently the odds would still be 128–1 if only 7 balls had been withdrawn and all were found to be white: or if on withdrawing 99, 53 were white and 46 black. Or even if 9999 balls had been taken out and 5003 found to be white.

Similarly, the calculations did not depend on the order in which the colours appeared. All 20 white followed by 13 black would have the same effect as a sequence in which the colours were mixed; or where 33 balls were emptied out and sorted into two piles which, on being counted, were of sizes 20 and 13.

The upshot of these two observations is that only the difference of numbers of white and black balls matters. We say the difference is *sufficient*. It is sufficient for you to do the Bayes' calculations and no other features of the data are relevant. It is an important task to recognize sufficient values because they can reduce a rather complicated situation to a simpler one.

However, the two observations also serve as a reminder for us not to forget H, the background information. In the case of 9999 withdrawn balls, 5003 white, the proportion of white balls observed is close to a half, which is not reasonable on the supposition in H that either 2/3 or 1/3 are white. (Notice we are here using statistical ideas or chances, rather than probabilities: see section 2.4.) In practice, one would revise H and suspect that the urn truly contained about equal numbers of balls of the two colours.

Similarly, if all the white balls appeared first followed by all the black ones, you might suspect that they were not being drawn at random (another part of H). The moral is never to forget H. Just how you change H is technically rather difficult and will not be discussed here. Of course, the technique uses only the rules of probability. An example of an 'urn' with many possible compositions is discussed in section 6.10.

6.6 INFORMATION-PROCESSING ON COMPUTERS

Bayes' theorem describes how we learn, or, more correctly, how we ought to learn. It says how our beliefs, expressed in terms of prior probabilities, should be modified by information described by the likelihood to give new beliefs expressed in the posterior probabilities. It does not describe all learning, for example, the acquisition of manual skills, but it does cover most adult learning. Alternatively the theorem can be regarded as a way of processing data. Modern computers, which are very fast at doing this, operate using Bayes' theorem. The following example illustrates this.

Suppose a patient is visiting his doctor for the first time. Then even before the patient has entered the room, the doctor will have some idea of what is the matter with him. Many of his patients have minor ailments, such as the common cold; a few will have more serious afflictions, such as rheumatism; but very few, at least in more advanced societies, will have deficiency diseases, such as beri-beri. These ideas can be expressed as prior probabilities of the patient having the various diseases, which are the uncertain events of the situation. When the patient enters he will describe his symptoms, the doctor will ask questions, carry out tests, and generally acquire information, X. The doctor has to process his prior beliefs and X to produce posterior beliefs about what is wrong with the patient.

Without going into details we can see how this works. If a disease θ_j is such that the set of symptoms X is rarely associated with it, $p(X \mid \theta_j$ and $H)$, the likelihood, will be small and therefore its product with the prior will also be small, so that the doctor will think it unlikely that the patient is suffering from θ_j. Also if the prior probability is small, even if X is often associated with θ_j, the product will be small and the doctor will not associate the disease with the patient. For example, he will not consider beri-beri until the symptoms are very emphatic. So a doctor broadly works according to the ideas in Bayes' theorem even if he has never heard of them.

What could easily happen in the not-too-distant future is that the doctor could have a console on his desk connected to a computer. He would type on the console the information X and the computer would be programmed to use Bayes' theorem and tell the doctor, through the console, what was most likely to be wrong with the patient. The prior values $p(\theta_j \mid H)$ could be stored in the computer, as could the likelihoods $p(X \mid \theta_j$ and $H)$. The former would come from experience of general practice, having relatively high values for common ailments and low ones for rare diseases. The latter would come from the whole of medical experience, describing how often patients, known to be suffering from θ_j, have developed symptoms X. The only additional information the computer needs is X, which the doctor supplies. Not only does this have the advantage that the relevant multiplications are done correctly, but it has the merit that the whole of medical experience and not just the doctor's own is used. The computer stores more information than the doctor and its data-processing is simply Bayes' theorem. Looked at as a tool to help the doctor, and not as a substitute for him, the idea is potentially very valuable. The following numerical example, confined for simplicity to three diseases, may assist the understanding.

Diseases	Prior probabilities $p(\theta_j \mid H)$	Likelihoods $p(X \mid \theta_j$ and $H)$	Prior times likelihood	Posterior probabilities $p(\theta_j \mid X$ and $H)$
θ_1	0.6	0.2	0.12	0.33
θ_2	0.3	0.6	0.18	0.50
θ_3	0.1	0.6	0.06	0.17
	1.0		0.36	1.00

One disease, θ_1, is relatively common; one, θ_3 is rare; the other, θ_2, is intermediate. The observed symptoms X rarely occur with θ_1 but are common with both θ_2 and θ_3. (Notice that the likelihoods of the θ_j, not being probabilities of θ_j, but of X, do not add to one.) The fourth column of the table gives the product required of Bayes' theorem (the right-hand side of equation (6.1)). These add to 0.36. Consequently $p(X \mid H)$ must be 0.36, and on dividing by this value we obtain the posterior probabilities in column five, which do add to one. The effect of the symptoms is to make the intermediate disease seem to be the most probable whereas the disease originally suspected, θ_1, has dropped to second place and the outside chance θ_3 has almost doubled in probability. The reader will learn much from varying the prior probabilities and/or likelihoods and recalculating the posterior values.

This is how all computers should process information. A new type of engineer has recently emerged: he is called a 'knowledge engineer'. There is nothing new about this. Bayes was the first knowledge engineer and all the modern one needs is Bayes' theorem. Sometimes this breed of engineer uses other ideas without recognizing their basic incoherence.

6.7 CROMWELL'S RULE

A simple result that follows from Bayes' theorem is that it is inadvisable to attach probabilites of zero to uncertain events, for if the prior probability is zero so is the posterior, whatever be the data. This is immediate since the latter is proportional to the product of the likelihood with the former, and a product is necessarily zero if one of its factors is. Consequently an uncertain event of zero probability remains so whatever information is provided. In other words, if a decision-maker thinks something cannot be true and interprets this to mean it has zero probability, he will never be influenced by *any* data, which is surely absurd. So leave a little probability for the moon being made of green cheese; it can be as small as 1 in a million, but have it there since otherwise an army of astronauts returning with samples of the said cheese will leave you unmoved. A probability of one is equally dangerous because then the probability of $\bar{E}$ will be zero. So never believe in anything absolutely, leave some room for doubt: as Oliver Cromwell told the Church of Scotland, 'I beseech you, in the bowels of Christ, think it possible you may be mistaken'. The only exception to these statements is when E follows logically from H, then $p(E \mid H) = 1$. Thus if H contains the rules of arithmetic, $p(2 \times 2 = 4 \mid H) = 1$.

We call this Cromwell's rule: that $p(E \mid H) < 1$ unless H logically implies E is true. Similarly, $p(E \mid H) > 0$ unless H implies E is false. For example, I believe the argument in this book is correct but have to reserve a small probability that it is wrong. Just how small this is depends on what is meant by 'wrong'. If wrong in comparison with other methods currently available, then it is very small, say 1 in 1000. But if wrong in the sense that ultimately other methods will supersede it, then it is nearly 1; for scientific experience shows that all ideas ultimately get modified. Thus Newton's laws are replaced by Einstein's: and Einstein's by?

6.8 INVESTMENT ADVICE

Another illustration of the use of Bayes' theorem is provided by the simple investment example of section 5.1. Consider the decreasingly risk-averse decision-maker with 1000 dollars assets (and a scale factor of 10 for his utility) contemplating the investment of 500 dollars in the stock which can win or lose him 100 dollars. Suppose that on reflection he feels that the stock is equally likely to appreciate (θ_1) or depreciate (θ_2), then because of his risk aversion he will not invest in the stock. He might feel it worth while to obtain expert advice on the subject before reaching a decision: after all it only needs the chance of appreciation, $p(\theta_1)$, to reach 52% to make the purchase attractive, so he consults his stockbroker who says the stock will appreciate and therefore advises purchase. How should the investor act; should he accept the broker's advice or not?

It clearly depends on how much he trusts his advisor; complete faith will undoubtedly make him risk his money, but suppose he has his doubts? Let us look at the problem with the aid of Bayes' theorem.

We have that $p(\theta_1 \mid H) = 1/2$ where H denotes the investor's knowledge. Let S denote the stockbroker's opinion that the stock will appreciate. We require $p(\theta_1 \mid H \text{ and } S)$, the chance of appreciation given both the advice and the investor's understanding of the situation. The additional feature needed is the likelihood $p(S \mid \theta_1 \text{ and } H)$. Rather roughly, the probability involved here is that of the broker being right in detecting a favourable investment, and it measures part of the faith the decision-maker has in his broker. That this is relevant is just the conclusion we came to in the preceding paragraph and all we have done is to confirm and quantify the idea. Suppose, for example, he respects the opinion to the extent that he assesses the broker's chance of being right at 0.8, then

$$p(\theta_1 \mid H \text{ and } S) \propto 1/2 \times 0.8 = 0.4$$

and similarly

$$p(\theta_2 \mid H \text{ and } S) \propto p(S \mid \theta_2 \text{ and } H)/2$$

The new probability here is the likelihood of the broker being wrong (giving the appreciation advice when the stock drops). If the decision-maker assesses this at 0.2,

$$p(\theta_2 \mid H \text{ and } S) \propto 1/2 \times 0.2 = 0.1$$

The two probabilities, for θ_1 and θ_2, must add to 1, so

$$p(\theta_1 \mid H \text{ and } S) = 0.8 \text{ and } p(\theta_2 \mid H \text{ and } S) = 0.2$$

The investor would now regard the risk as worth taking. The expected utility is $0.709 \times 0.8 + 0.676 \times 0.2 = 0.702$, greater than the utility, 0.693, of leaving the money in the bank. Notice that the probability of appreciation, 0.8, is substantially greater than the critical probability, 0.52, so that even if the investor's assessment of his stockbroker's abilities is optimistic, a revised figure would still suggest investment. In fact, even if the stockbroker is only a little bit better than a person who advises by tossing a fair coin, and has $p(S \mid \theta_1 \text{ and } H) = 0.52$ and $p(S \mid \theta_2 \text{ and } H) = 0.48$, calculations show that $p(\theta_1 \mid S \text{ and } H) = 0.52$ and investment is just worth considering. The example illustrates the way advice can be used, but does not take into account the cost of the advice. This is a topic we turn to in the next chapter.

In the example we supposed $p(S \mid \theta_1 \text{ and } H) = 1 - p(S \mid \theta_2 \text{ and } H)$, interpreting the former as the chance of the broker being right and the latter as his chance of being wrong. However, the interpretation is not quite correct for the broker may be better at detecting a rising stock (θ_1) than a falling one (θ_2), and he is said to be *biased*. Another example serves to illustrate the point.

6.9 ERRORS OF THE TWO KINDS

Let us return to the mountain-pass example of section 3.8. Let θ_1 be the event that the pass is blocked and θ_2 the contrary event that it is open. Suppose again $p(\theta_1 \mid H) = 1/2$. Now being so uncertain, it might be an idea to invest in a telephone call to one of the motoring organizations and ask them whether the pass is free. Suppose they say it is blocked and denote this by S. Then the analysis goes as before and we require $p(S \mid \theta_1 \text{ and } H)$ and $p(S \mid \theta_2 \text{ and } H)$. My personal assessment of the probability of them saying the pass is blocked, when it really is, is 0.8—not 1, because I know news of a snowfall takes a while to reach them. Hence $p(S \mid \theta_1 \text{ and } H) = 0.8$ and we need $p(S \mid \theta_2 \text{ and } H)$. Here my personal assessment is not $1 - 0.8 = 0.2$ but more like 0.5, because I believe the motoring organizations prefer to be cautious in their advice with the object of keeping the motorist off the road—quite sensibly from their point of view. I believe they will often be wrong in reporting the pass blocked when it is free, but will usually be right when it is truly blocked. So, for me,

$$p(\theta_1 \mid H \text{ and } S) \propto 1/2 \times 0.8 = 0.4$$

but

$$p(\theta_2 \mid H \text{ and } S) \propto 1/2 \times 0.5 = 0.25$$

This gives the chance of the pass being blocked, on receipt of advice that it is, as $0.4/0.65 = 0.62$, not 0.8 as with the stockbroker. The probabilities express quantitatively my personal opinion that the organization's information is biased.

So we see that in assessing the value of advice we need two quantities, $p(S \mid \theta_1)$ and $p(S \mid \theta_2)$, omitting reference to H. Alternatively, $p(\bar{S} \mid \theta_1)$ can replace $p(S \mid \theta_1)$ since they add to 1. Now both $p(\bar{S} \mid \theta_1)$ and $p(S \mid \theta_2)$ correspond to bad advice, to errors. The first is advising no appreciation when the stock rises: the second, appreciation when it does not. The mountain example shows that these can be different. There they were 0.2 and 0.5, respectively. They are the errors of the two kinds and both need to be specified in the coherent approach.

6.10 ATTRIBUTE SAMPLING

We next consider a situation that often arises. Suppose that there is a large collection of things, some of which are of one type, the rest of another, and it is desired to know the proportion that is of the first type. Suppose the type is distinguished by the first having an *attribute* that the second does not. With balls in the urn, the attribute might be black colour. To provide information about this some of them are taken at random and their types found. This is called *sampling*. How is the uncertainty about the proportion changed by this evidence? We have already met some examples: balls in urns which are black or white, material which is defective or sound. Others easily spring to mind: voters who will support a particular party in a two-party election, patients who will respond favourably to a drug. To assist understanding, we will work in terms of a specific situation, namely that of the producer of curtain material discussed first in section 4.1.

Suppose that the rolls of material are produced in batches of 1000 or more. The manufacturer knows that there is a lot of variation between batches; some being almost free of defects, others having large numbers, for reasons that he either does not understand or finds too expensive to avoid. Then a sensible thing for him to do would be to take some of the rolls from a batch at random, inspect them, and if the number found is small to accept the batch, otherwise to reject it. In this way he will learn about the overall quality of the batch and hence about individual rolls in the batch. Specifically, suppose he inspects n rolls and finds r defective. In a numerical example considered below $n = 10$ and $r = 1$. The problem is similar to sampling balls from an urn (section 6.3) except that there only two possible proportions were assumed, 1/3 and 2/3. Here, in a batch of 1000, there are 1001 possibilities. Denote the possible values by θ_j. Then by Bayes' theorem

$$p(\theta_j \mid r, n) \propto p(r, n \mid \theta_j) p(\theta_j)$$

Here X, the information, is the pair (r, n) and, as usual, reference to H is omitted. As before, the prior information $p(\theta_j)$ about the batch quality is changed by the information (r, n) to produce the posterior opinion $p(\theta_j \mid r, n)$. The change is effected by multiplying by the likelihood $p(r, n \mid \theta_j)$ and then standardizing so that the results add to 1.

We first calculate the likelihood. If the proportion of defectives is θ_j, or the chance is θ_j, the probability that the first roll will, on inspection, be found defective is also θ_j (section 2.4); and the probability of being satisfactory is $1 - \theta_j$. The same will hold for the second roll if the batch size is large since the removal of one roll will scarcely alter θ_j; and for all subsequent rolls if the total number sampled n is small in comparison with the batch. By the multiplication law (section 3.14) these probabilities may be multiplied to give that for r defects out of n. Hence

$$p(r, n \mid \theta_j) = \theta_j^r (1 - \theta_j)^{n-r} \tag{6.2}$$

is the likelihood. Bayes' theorem now reads

$$p(\theta_j \mid r, n) \propto \theta_j^r (1 - \theta_j)^{n-r} p(\theta_j)$$

Let us insert some numerical values to bring the problem alive. The values

Table 6.1

θ_j	$p(\theta_j)$	$\theta_j(1-\theta_j)^9$	Product	$p(\theta_j \mid r=1, n=10)$
0.00	0.10	0.0000	0.0000	0.00
.10	.20	.0387	.0078	.36
.20	.40	.0268	.0107	.50
.30	.20	.0121	.0024	.11
.40	.10	.0040	.0004	.02
	1.00		0.0213	1.00

of θ_j present a difficulty, for if the batch is of size 1000 there are 1001 of them. The point can be surmounted using the calculus but in default of this let us suppose θ_j restricted to five values; namely 0.0, 0.1, 0.2, 0.3, and 0.4. This is artificial but will simplify the arithmetic without destroying the essence of the problem. The calculations are given in Table 6.1.

The first column lists the values of θ_j, the possible proportions. The second provides the prior probabilities. For example, the first entry says there is a 10% probability of there being no defectives. (These values are subject to a coherence requirement discussed in the next section.) For a sample of 10 with exactly 1 defective, the third column displays the likelihood and the fourth gives the product of prior and likelihood that occurs in Bayes' theorem. These products add to 0.0213, and on division of each by this amount we get the posterior probabilities listed in the final column.

In commentary on the table, the original opinion is that the proportion defective is most likely around 20% but could be as high as 40% or as low as zero. The likelihood values show that the observed single defect in the sample of 10 is most likely when the batch proportion is 10% but is still fairly likely when it rises to 20% or even 30%. It is, of course, impossible if $\theta_j = 0$ and there are no defects. The final opinion is still that 20% is the most probable proportion (at 0.50) but 10% is probable (0.36), and 30% is not ruled out (at 0.11).

The calculations are repeated in Table 6.2 for a sample of 20 rolls inspected of which 2 were found defective: $r = 2$, $n = 20$, still with 10% seen to be faulty. The final result is that there are essentially only two possibilities left amongst

Table 6.2

θ_j	$p(\theta_j)$	$\theta_j^2(1-\theta_j)^{18}$	Product	$p(\theta_j \mid r=2, n=20)$
0.00	0.10	0.00000	0.000000	0.00
.10	.20	.00150	.000300	.48
.20	.40	.00072	.000288	.47
.30	.20	.00015	.000029	.05
.40	.10	.00002	.000002	.00
	1.00		0.000619	1.00

the five originally considered, θ_j of 10 or 20%, and these are about equally probable. We see how the initial opinion (first column) changes after 10 rolls have been seen (last column of Table 6.1) and again with a further 10. No-one could see just how the values would change without using Bayes' theorem, and most people are surprised by how strongly the value 20% can be held (probability almost 1/2) despite only seeing 10% in the samples. The inspection can be thought of as organized learning about the proportion defective; the organization being Bayes' theorem.

6.11 PROBABILITIES FOR A SINGLE UNIT

The discussion of attribute sampling in the last section referred to the proportion of defectives in the batch, or generally to the proportion of one type in the things under consideration. This is not usually the final quantity of interest. The doctor does not want to know the proportion of patients who respond favourably to a drug: he wants to know whether the individual he is treating will respond favourably. Certainly, if you are that individual patient, that will be your concern. The way the curtain material problem was originally propounded (section 4.1) and then solved (section 4.4) was in terms of the probability of a single roll being defective, not of the proportion in a batch. How are these two ideas connected? In the last section $p(\theta_j \mid r, n)$ was calculated. We require $p(D \mid r, n)$, where D is the event that a single roll is defective. Extend the conversation from D to include θ_j. In section 3.8 the theorem was written

$$p(A) = \sum_{j=1}^{n} p(A \mid E_j)p(E_j)$$

Replace A by D, E_j by θ_j and let the conditions r, n (and H) be omitted from the notation. Then

$$p(D) = \sum_{j=1}^{n} p(D \mid \theta_j)p(\theta_j)$$

But if the proportion defective is θ_j and the roll is taken at random from the batch, the probability of any one roll being defective is also θ_j: hence $p(D \mid \theta_j) = \theta_j$ and so

$$p(D) = \sum_{j=1}^{n} \theta_j p(\theta_j) \tag{6.3}$$

This establishes a connection, or a coherence, between $p(D)$ and $p(\theta_j)$. In section 4.2 we had $p(D)$ (there $p(\theta_2)$) equal to 0.2. This was before sampling, given H only. The reader can verify that this agrees with the prior probabilities for θ_j listed in the first column of Table 6.1. Applying the same result to the posterior values in the same table we see $p(D \mid r = 1, n = 10) = 0.18$, and for those in Table 6.2, $p(D \mid r = 2, n = 20) = 0.16$. Hence the effect of the sampling is to reduce the probability of a single roll being defective because the sample values, 10%, are less than the original values, around 20%. In the decision problem (section 4.4) with the original probabilities, the expected utility of inspecting (d_1) was 0.82, of not, 0.80 and inspection of that batch was called for. After $r = 1$, $n = 10$, these utilities become 0.83 and 0.82, still with a slight, but reduced, advantage in inspecting. After $r = 2$, $n = 20$ they are 0.84 for both actions and neither is obviously superior to the other. We saw in section 4.4 that if $p(D)$ dropped to 0.1 then it is clearly better not to inspect, reversing the decision at $p(D) = 0.2$. The critical value of $p(D)$ that makes the two courses of action equally attractive, having the same expected utility, is 1/6, or about 0.17.

6.12 OTHER APPLICATIONS

The argument just given is really rather general and applies in many situations. Let us consider the basic ingredients. We have a large number of things, each of which either has, or has not, some well-defined property. In the example a batch of rolls, each of which is either good or bad. Some of them are inspected to see whether they have the property or not. We infer from this whether another one of them will have the property. Some further examples will help to emphasize how often a situation fits into this framework.

1. Coin tossing

A coin either falls heads or tails when tossed. It is tossed 10 times and a head appears only once. What would you infer about the result of the eleventh toss? There the equivalent of the batch is the imaginary possible tosses of the coin. A device, which will not be explored here, enables one to avoid the introduction of imaginary tosses whilst still preserving the algebra of the original examples.

2. Psychological testing

A subject is given a battery of tests, each of which he either passes or fails, with a view to employment. What is the chance that he will pass an additional test? This can be extended to answer what is the chance that he will be a suitable person to employ, on the grounds that the tests are indicators of his ability in more complicated situations.

3. Scientific experimentation

Our original example is typical of many that arise in the control of quality in industry. This next example is typical of much scientific practice. A pharmacologist has developed a new drug for a certain illness and wishes to try it out. To do this he selects several pairs of patients suffering from the illness and matches the members of a pair. Thus a pair might consist of two elderly women, otherwise healthy; another pair might be two young males with abnormal blood pressure. The point is that the two members of the pair are similar in respects likely to be relevant to the effect of the drug. The pharmacologist then gives one member of the pair his new drug, and the other member the standard drug used in the treatment of the illness. Then for each pair he will either obtain the result that the new drug is better, or the contrary, ignoring the possibility of ties. Just as the material is good or bad, the new drug is better or worse. From tests on several pairs he will be able to use the above argument to infer the chance of the new drug being better than the old when applied elsewhere. In other words, he will be able to assess the merit of the new drug over the old.

4. Forecasting tomorrow

This example is included to emphasize that the method just described makes some assumptions which must not be forgotten. You go on your holiday and out of the first 10 days you get only 1 dry one: what is the chance that it will rain tomorrow, the eleventh day? Although all the ingredients are there, the above analysis is inappropriate. The reason is that in the course of the discussion with the rolls of material (and with the balls from the urn) we assumed in the argument leading to the likelihood (equation (6.2)) that the chance of one roll being good does not depend on the other rolls. This is well known not to be true of weather; the chance of one day being wet does depend on whether the preceding day was wet. Equation (6.2) is therefore incorrect, and a more sophisticated evaluation incorporating the dependence of one day on the next is required. One can see that the argument will not work by another consideration. The calculation in the industrial example involved only r, the number of bad rolls, and not all the information. For example, with $n = 10$, $r = 1$ it was irrelevant whether the defective roll was the first, third, or last to be inspected. (With the urn example it was similarly noted that the order was irrelevant.) This is not true of weather. If you told me that you had had only one fine day out of the last 10, I would want to know whether that one day was today or not. If it was today, the chance of fine weather tomorrow would be higher than otherwise, since consecutive days tend to be alike, at least in most climates.

6.13 ALL UNCERTAINTY IS OF THE SAME TYPE

These examples will give some idea of the range of situations where the analysis is appropriate. They are also useful in attempting to dispose of an objection sometimes raised against our basic assumption in section 2.1 that there is only one kind of uncertainty in the world. The objection is related to that raised against the use of one number, expected utility, to reach a decision, an objection that was discussed and, we hope, disposed of in the last chapter. It is related because the objector feels that one number alone is insufficient to describe his probability assessments.

The objection runs as follows. There are, it is said, some probabilities that one can assess easily and feel reasonably sure about, but there are others that are terribly vague and it is doubtful whether the single number means anything, the objector only agreeing to give a number because the argument in this book has some conviction. How, the objection continues, can these two different things be described by a single number, surely the vagueness also needs to be specified? Some objectors have gone so far as to use upper and lower probabilities. These are close together if the original probability is firm and wide apart if vague. An example will illustrate the difficulty and also provide a basis for the reply.

Contrast example (1) above of coin tossing with the scientific example (3) of testing a new drug. In default of any data on the particular coin beyond an inspection of it to see that it was similar to coins one had experience of in the past, one would say that the probability of heads on a fair toss is 1/2, and this probability is firm in that we would almost all be happy with it. With the drug test one may have an open mind about its effectiveness and similarly ascribe a probability of 1/2 to the new drug doing better than the old with a particular pair of patients. The latter value of 1/2 is vague and one does not feel so sure about it as one does with the coin. There is not the experience of drug testing that there is with coins. So here we have one probability of 1/2 that is firm—some would say 'objective'—and another that is vague and definitely 'subjective'. Surely, the argument goes, they are not the same, as the theory of this book would have us believe.

The reply uses the device of extending the conversation in the same way that it was used in the industrial quality control illustration, there from a single roll to a batch. Instead of considering just the probability of heads on a single toss we introduce the idea of the proportion of heads in a batch of tosses. Instead of only the chance of the new drug succeeding with one pair, we investigate the proportion of a large number of pairs of patients in which the new drug does better. The idea of extension in this way is in agreement with our basic requirement of coherence between different assessments. The judgement on one occasion has got to be coherent with our judgements over a large number of similar occasions.

Imagine the coin tossed 1000 times and denote by θ_j, as before, the proportion of heads in these 1000 tosses. Here j runs from 1 to 1001 as θ_j runs from 0.000, 0.001 to 0.999 and 1.000. Consider, as we did above, $p(\theta_j \mid H)$, the prior probability of θ_j, and also $p(D \mid H)$, the probability of the event D of heads on a single toss. (We use the same notation as for a single roll of material being defective.) These probabilities are related in the form described by equation (6.3). Consider what values it is reasonable to give to $p(\theta_j \mid H)$. Most people would ascribe very low values to it when θ_j was near zero or 1, for they would hardly expect even as few as 200 heads, or as many as 800, in a thousand tosses. The only values with reasonable probabilities would be around $\theta_j = 1/2$.

In the drug illustration denote by ψ_j the proportion of 1000 pairs of patients benefiting more from the new drug than the old. (There ψ is another Greek letter, 'psi'.) Now ψ_j has the same range of values as θ_j but the probabilities

$p(\psi_j \mid H)$ are likely to be somewhat different from those associated with the coin-tossing. Thus if we consider a low value of ψ_j near zero, this will correspond to the case where the new drug is a complete flop and the old cure is vastly to be preferred. This may have a small probability, but it is certainly a possibility which must be seriously considered. The same remark applies to values of ψ_j near 1, where the new drug is superior. Both low and high values of ψ_j are much more probable than low or high values of θ_j, simply because we know that coins could not easily be that biased, whereas drugs could well be that different. What we are saying here is that although $p(D \mid H)$ may be the same in the two circumstances, the values of $p(\psi_j \mid H)$ and $p(\theta_j \mid H)$ are quite different, for the former all the values are appreciable, whereas in the latter only those referring to θ_j around 1/2 are of any magnitude. These different probabilities reflect the vagueness and firmness that are respectively associated in our minds with the original equal probabilities.

6.14 THE EFFECT OF DATA ON 'VAGUE' AND 'PRECISE' PROBABILITIES

This vagueness and firmness can also be appreciated by the effect data will have on the probabilities. Suppose, on the one hand, 12 tosses of the coin all result in heads, and on the other, of 12 pairs of patients all do better with the new drug. Most people's intuitive reactions in these situations will be quite different. With the coin they will probably shrug their shoulders and say we have just had an extraordinary result and remain convinced (though perhaps not quite as strongly as before) that the coin is reasonably fair. On the other hand, the medical evidence is quite likely to lead them to believe fairly strongly in the effectiveness of the new drug. The firm probability has shifted but little, whereas the vague one has changed a lot. The chance that the thirteenth toss (ignoring superstitions) will be heads is still close to 1/2: the chance that the new drug will do better with the thirteenth pair of patients is much nearer to 1. These intuitive ideas are supported by Bayes' theorem as will now be demonstrated.

Denote the data by X: then

$$p(\theta_j \mid X \text{ and } H) \propto p(X \mid \theta_j \text{ and } H)p(\theta_j \mid H)$$

as before, with a similar result with ψ for θ. The data just considered, of 12 experiments all giving the same results, are such that $p(X \mid \theta_j \text{ and } H)$ and $p(X \mid \psi_j \text{ and } H)$, the two likelihoods, are only appreciable for θ_j and ψ_j near 1. (In fact, their values are θ_j^{12} and ψ_j^{12}, respectively.) What is going to be the effect on Bayes' result? With the coins $p(\theta_j \mid H)$ is so small except around $\theta_j = 1/2$ that the product with the likelihood, even at its largest value around $\theta_j = 1$, will be small, again except near $\theta_j = 1/2$. Consequently one remains fairly sure that θ_j is near 1/2. The opposite happens with the patients, $p(\psi_j \mid H)$ is of much the same order of magnitude everywhere and consequently its product with the likelihood will only be appreciable where the latter is, that is, around

$\psi_j = 1$. Thus one becomes much more convinced that ψ_j is near 1. These changes are reflected in similar changes in $p(D \mid H)$ by equation (6.3).

We therefore claim that a single number, a probability, is adequate for decision-making. Associated with it may be other ideas of vagueness or firmness. These ideas are irrelevant for the decision in hand, but are relevant, and are accurately reflected in the laws of probability, when additional data are available. Exercise 6.7 at the end of this chapter is designed to help the reader to more understanding of the situation. In the following paragraph a technical device is described which may clarify the point still further.

6.15 SIMPLIFIED CALCULATIONS

In many analyses it is possible to describe the whole set of the probabilities $p(\theta_j \mid H)$ (of which, in the example, there were 1001) by two values. Just how this can be done is something that would require a lot of technical material to describe, and is therefore omitted. Let these two values be denoted by s and m; s is not greater than m, and typically both are integers and both are positive. These values have the following meanings: their ratio s/m is the value of $p(D \mid H)$, the chance for a single occasion, and m measures the firmness associated with that probability, higher values corresponding to more conviction about the value of $p(D \mid H)$. Thus with the coins we might have $s = 1000$, $m = 2000$, giving a ratio of 1/2 and a large value of m corresponding to a belief that the coin, if biased, is only affected by a small amount. With the drug, on the other hand, we could use $s = 1$, $m = 2$, still with a ratio of 1/2 but the low value of m expressing the vagueness of our ideas.

The effect of data on s and m is simple. For every observation m increases by 1, but s only increases by 1 if the event whose probability is under discussion occurs, otherwise it remains unaltered. The effect on the ratio is therefore a shift either to $s/(m+1)$ or $(s+1)/(m+1)$, a shift which is small if m is large, but appreciable when m is small. Thus with the 12 tosses, all of which resulted in heads, the ratio changes from 1000/2000 to $1012/2012 = 0.503$, a slight change. Whereas in the case of the 12 pairs of patients all benefiting by the new drug, the ratio changes from 1/2 to $13/14 = 0.929$, a substantial alteration. This demonstrates a quantitative description of the vagueness by measuring the response of the probability to additional data. Notice that m necessarily increases so that evidence can only make probabilities firmer in this sense.

There is much more that might be said about the ways in which data affect beliefs but any such discussion would tend to involve technicalities which we wish to avoid. We therefore now turn our attention to the question of whether the data should be collected and ask if it is worth having. This is the topic of the next chapter.

Exercises

6.1. A certain phenomenon can either be present E, or absent, $\bar{E}$. An apparatus has been designed to detect the phenomenon but it is not infallible. If the phenomenon is present, the apparatus has chance p_1 of detecting it by giving a positive response, but even if it is not, the apparatus has chance p_2 of giving a positive response. (p_2 is less than p_1.) If the apparatus has just given a positive response, find the probability that the phenomenon is present in terms of p_1, p_2, and $p(E)$.

Apply the result in the following situations:

(a) the apparatus detects whether an airline passenger is carrying explosives, $p(E)$ is about 1 in a million. The apparatus always detects explosives, $p_1 = 1$, but gives about 1% of false alarms, $p_2 = 0.01$. If the apparatus has responded what is the chance that the passenger is carrying explosives?

(b) Patients are suspected of having a certain disease, E. A standard test has $p_1 = 0.8$ and $p_2 = 0.4$. If 50% of suspects have the disease what is the chance that a tested suspect who responds positively has got the disease? What is the same chance when the response is negative?

6.2. (This question is a simplified form of the situation that arises in countries that have a law relating car-driving and alcohol content of the blood.) A quick police test of the alcohol content has only a chance 0.8 of being correct: that is, of giving a positive response when the content is high or of a negative value when it is low. Suspects, that is, those with a positive reading, are given a careful test by a doctor. This test never makes an error with a sober driver but does have a 10% chance of error with an inebriated one because of the time-lapse between the two tests. The two tests may be supposed independent. When a driver is stopped by the police the chance of his having a high alcohol content is P.

(i) What proportion of drivers stopped by the police will have a second test which does not detect alcohol?

(ii) What is the posterior probability that such a person did, in fact, have a high alcohol content?

(iii) What proportion of people given a police test will not have a second test?

6.3. A patient thinks he may have cancer and consults his doctor, A, who, after examination, declares he has not. The patient feels his doctor is overcautious about diagnosing cancer so he consults a second doctor, B, who declares he has cancer.

Suppose doctor A diagnoses cancer in only 60% of those patients who have it, and never in the case of those who do not. Suppose B diagnoses cancer in 80% of those patients who have it and in 10% of those who do not. Suppose the two opinions are given independently.

By what factor are the patient's odds against having cancer multiplied as a result of the two opinions?

6.4. In a multiple-choice question an examinee is given a choice between m answers. Suppose there are only two possibilities: either he knows the answer, with probability p, or he guesses, selecting one of the answers with probability $1/m$. If an examinee has answered a question correctly what is the chance that he really knew the answer and was not guessing?

6.5. (This is a classic problem due to Bertrand.) There are three closed boxes and you select one at random. It is known that one box contains 2 gold coins, one contains 2 silver coins, and the remaining box contains 1 gold and 1 silver coin. The first coin taken from the selected box is gold, what is the chance that the second coin will also be gold?

6.6. (This question arose in a law case in California.) Witnesses reliably testified that the crime had been committed by a couple consisting of a negro man with a beard and moustache and a Caucasian girl with blonde hair and a ponytail, driving a partly yellow automobile. The defendants had all these characteristics but were not clearly identified as the perpetrators of the crime. The chance that any couple possesses all the above characteristics is one in 12 million. What can you say about the chance that the defendants are guilty?

6.7. (This question is designed to assist the comparison of 'firm' and 'subjective' probabilities and uses the notation in the chapter.) Suppose ϕ_j and ψ_j can take any of the eleven values 0.0, 0.1, 0.2, ... 1.0. In the coin-tossing case suppose $p(\phi = 0.5) = 0.9$ and $p(\phi = \phi_j) = 0.01$ for $\phi_j \neq 0.5$. In the drug-testing situation suppose $p(\psi = \psi_j) = 1/11$ for all j. Suppose 5 heads are observed in 5 tosses, and 5 times out of 5, drug A is better. Calculate the respective posterior probabilities on the above evidence.

(The fifth powers of the first nine integers are 1, 32, 243, 1024, 3125, 7776, 16807, 32768, and 59049.)

Chapter 7

Value of Information

'"I don't know," said Soames. "We are here to decide policy according to our common sense, and we must have the fullest opportunity of exercising it. That is my point. We have not enough information."'

The White Monkey, Ch. 7.

7.1 PERFECT INFORMATION

In the last chapter we studied the effect data have on the probabilities of the uncertain events in a decision problem. In this chapter we evaluate the improvement in decision-making that results from these revised probabilities. We speak of the data providing information, and show how this information can be measured. As a result of the evaluation it becomes possible to assess the expected gain from the data before they are available, and therefore to see whether the data are likely to be worth having.

We first consider the situation where the data are completely informative, in the sense that they tell us exactly which uncertain event obtains. This is described as acquiring *perfect information*. Only after having discussed this case do we pass to the partial information situation which occupied our attention in the last chapter, where the data modify the probabilities but still leave some uncertainty in the problem. The restriction to the case of perfect information is not the only simplification to be made in the earlier discussion, it will also be supposed that the consequences are, as in Chapter 5, entirely monetary, and moreover that over the range of money values being considered in the decision problem, the utility of money is effectively linear. In section 5.12 we saw that this would typically be true if the range was small in comparison with the total assets; an example cited was that of an insurance company considering a single policy. The argument used in this very restricted situation will then be generalized to deal with other cases.

7.2 INVESTMENT EXAMPLE

We begin with the investment example discussed in previous chapters (for example, Table 5.1) in the case where the decision-maker's assets are large in comparison with the 100 dollars he might win or lose with the investment. Specifically, suppose his assets are 5000 dollars so that the decision table is as follows, with the consequences described in dollars:

	θ_1: Stock appreciates	θ_2: Stock depreciates
d_1: Invest	5100	4900
d_2: Leave in bank	5000	5000

(The reader may like to take either of our reference decision-makers I or II, with a scale factor of 100, say, and verify that for them utility of money is almost proportional to money over the range from 4900 to 5100 dollars.)

The problem to be considered is this: if the decision-maker knew of some way of finding out for certain how the stock was going to behave, how much should he be prepared to pay for this perfect information before receiving it? Suppose, for example, he had a perfectly accurate stockbroker to advise him; what is a reasonable fee for him to pay the broker? The case of the informative, but sometimes inaccurate, adviser considered in the previous chapter will be discussed later. Clearly, the value of the advice will depend on the probability that the stock will appreciate; for, to take an extreme case, if one was fairly sure that it would, there is little to be gained by advice. Suppose first that this probability is 1/2: in symbols $p(\theta_1) = p(\theta_2) = 1/2$. It will be assumed that the fee is paid before the perfect information is available, or at least that the fee is fixed, irrespective of the data. The simpler case of payment by results will be mentioned later.*

The perfect information provided can take two forms: either the decision-maker is told that the stock will appreciate (θ_1 is true) or that it will depreciate (θ_2 is true). If θ_1 is true, d_1 is the best decision and the monetary outcome will be 5100 dollars. If θ_2 is true, d_2 is best and he will have 5000 dollars. Consequently, with perfect information the decision-maker will have either 5000 or 5100 dollars. The difficulty is to know which. This we cannot say because the perfect information depends on θ_1 and θ_2 and we are uncertain about these when the fee is paid. However, we do have probabilities for them, namely 1/2 and 1/2. Consequently the *expected* outcome with perfect information is

$$1/2 \times 5100 + 1/2 \times 5000 = 5050$$

Without the information the expected utility is 5000 dollars (given by either decision), so that the expected value of the perfect information is the difference, 50 dollars. It would be worth paying up to 50 dollars for this unassailable advice.

An alternative way of looking at the situation is to introduce explicitly a fee,

* A stockbroker's fee is usually a commission on the purchase of the stock; that is, it only arises if d_1 is selected. This may bias his advice! See Exercise 7.4.

f, for the advice. Then after paying the fee but before receiving the information the decision table is as follows:

	θ_1	θ_2
d_1	$5100 - f$	$4900 - f$
d_2	$5000 - f$	$5000 - f$

Then if the advice is θ_1, d_1 is selected and $5100 - f$ obtained: if θ_2, d_2 and $5000 - f$ obtained. The expectation is $5050 - f$, which equals the 5000 available without information if $f = 50$. This approach will be of value when discussing more general utilities.

The effect of different opinions about the uncertain events can be seen by considering other values of $p(\theta_1)$. Thus if $p(\theta_1) = 3/4$ and $p(\theta_2) = 1/4$, so that one is fairly confident that the stock will appreciate, then without the advice it is best to invest (d_1) with expectation 5050 dollars. If the advice is for θ_1, then d_1 is optimum with 5100; and if for θ_2, then d_2 yields 5000, as before. But these now have probabilities 3/4, 1/4 respectively, so that the expectation with perfect information is 5075 dollars. The difference, 25 dollars, is the expected value of the information. Notice that this is only one half its value when one had equal probabilities for θ_1 and θ_2. The reader may like to try the case $p(\theta_1) = 1/4$, $p(\theta_2) = 3/4$ where one is dubious about the stock. Because of the symmetries in the problem this again gives an expected value of perfect information of 25 dollars. This contradicts some people's intuitive feeling that it is more valuable to seek advice about promising investments than about doubtful ones.

7.3 EXPECTED VALUE OF PERFECT INFORMATION

Let us now pass from this special decision problem and try to analyse a general decision problem in the same way. Such a problem is illustrated in Table 7.1 for the case of three decisions and four events, though some simplification of notation has been introduced. (The reader may like to compare Table 4.2.) The decisions, d_i, and the uncertain events, θ_j, remain unaltered but the utilities have been written u_{ij} instead of $u(C_{ij})$ and the probabilities $p(\theta_j)$ become simply p_j. To calculate the expected utility associated with any decision d_i one takes the utilities in the row corresponding to that decision, multiplies each by its associated probability (given in the table at the foot of the column), and sums the results. In symbols, the expected utility of d_i is

$$\sum_{j=1}^{n} u_{ij} p_j \tag{7.1}$$

(compare equation (4.2)). The best decision is that for which this expression is a maximum. In other words, the expected utility to be obtained without

Table 7.1. Decision table

	θ_1	θ_2	θ_3	θ_4
d_1	u_{11}	u_{12}	u_{13}	u_{14}
d_2	u_{21}	u_{22}	u_{23}	u_{24}
d_3	u_{31}	u_{32}	u_{33}	u_{34}
Probabilities	p_1	p_2	p_3	p_4

additional information is equal to the expression (7.1) for one value of i, namely that corresponding to the row that maximizes it.

Now suppose there is perfect information. If the uncertain event is θ_j one selects the decision which gives greatest utility in the column corresponding to that uncertain event. The expected utility with perfect information is obtained by multiplying this maximal utility by the corresponding probability of that uncertain event and summing all the products obtained for the different events. In symbols the expected utility with perfect information is

$$\sum_{j=1}^{n} \max_{i} u_{ij} p_j \tag{7.2}$$

The difference between this quantity (7.2) and the maximum value of expression (7.1), that is

$$\sum_{j=1}^{n} \max_{i} u_{ij} p_j - \max_{i} \sum_{j=1}^{n} u_{ij} p_j$$

is called the *expected value of perfect information*. Strictly, it is the expected gain in expected utility that results from the perfect information. Before the information is available, it is a reasonable assessment of what value the information will provide.

7.4 EXPECTATION AND REALIZATION

It is worth noting that it is reasonable to describe it as a gain since it is never negative, or, expressed differently, one can always expect perfect information to be worth having. The proof is immediate since whatever value of i is chosen in expression (7.1), that is, whatever d_i is selected without perfect information, expression (7.1) cannot exceed formula (7.2), since every utility, u_{ij}, in the former expression is replaced by a utility,

$$\max_{i} u_{ij},$$

in the latter, which is certainly not smaller. Hence this is true even if the d_i which maximizes formula (7.1) is selected, which proves the result.

It is also easy to see when the two expected utilities are equal and therefore the expected value is zero. For suppose d_s is the optimum decision without perfect information, so that formula (7.1) is maximized for $i=s$. The expected value is

$$\sum_{j=1}^{n} \max_i u_{ij}p_j - \sum_{j=1}^{n} u_{sj}p_j$$

and this can only be zero if

$$\max_i u_{ij} = u_{sj}$$

for every j. In other words, d_s must be such that it is the best decision for every θ_j. A decision which is best for every uncertain event clearly needs no information about which event obtains, for the knowledge would be superfluous. Apart from this trivial case, there is always a positive gain to be expected from perfect information.

Despite this it is well to remember that although the gain is *expected*, it need not always be realized. To illustrate this consider the following decision table with utilities proportional to the stated dollars:

	θ_1	θ_2
d_1	4800	5200
d_2	5000	5100
pr	1/2	1/2

Without perfect information d_2 is the better decision with an expectation of 5050 dollars. With perfect information the expectation is

$$1/2 \times 5000 + 1/2 \times 5200 = 5100$$

and the expected value of perfect information is 50 dollars. But when the perfect information is that θ_1 obtains, d_2 is selected and the return is 5000 dollars, less by 50 dollars than the expected return without any information of 5050. Consequently the information can reduce one's expectation. Information can be good or bad, but in the sense in which the term is used here, it is always expected to be good.

7.5 LOSSES

It is now easy to obtain a useful alternative expression for the expected value and thereby to provide another way of looking at a decision problem. Suppose that from each utility u_{ij} in Table 7.1 a number, let us call it a_j, is subtracted which depends on j, referring to the uncertain event, but not on i, referring to the decision. That is, all the utilities corresponding to the same uncertain event (and therefore in the same column) have their values reduced by the same amount, but the amounts may vary with the uncertain events. Then it is clear that, whatever be i, expression (7.1) is reduced by the amount,

$$\sum_{j=1}^{n} a_j p_j$$

and also that expression (7.2) is reduced by the same amount, since

$$\max_i u_{ij}$$

the greatest utility in the column corresponding to θ_j, will also be reduced by a_j if every utility is. Consequently both expressions (7.1) and (7.2) are reduced by the same amount, and the expected value, which is their difference, is unaffected. It remains only to take

$$a_j = \max_i u_{ij}$$

that is, to reduce the largest utility in each column to zero, to obtain a value of zero for expression (7.2), and hence the expected value, which is the difference between formulae (7.2) and (7.1), to be zero minus

$$\max_i \sum_{j=1}^{n} (u_{ij} - a_j) p_j$$

with $a_j = \max_i u_{ij}$. Define

$$l_{ij} = \max_i u_{ij} - u_{ij} \tag{7.3}$$

then l_{ij} is the difference between the utility of the best decision for θ_j and the utility for d_i when θ_j obtains. It is called the *loss* in deciding on d_i when θ_j obtains, since it measures how far from the optimum d_i is when θ_j is true. Since changing the sign turns a maximum into a minimum, we finally obtain for the expected value

$$\min_i \sum_{j=1}^{n} l_{ij} p_j \tag{7.4}$$

In words, the result says that the expected value is equal to the least expected loss obtainable with any decision.

Let us verify this with the investment example. The losses are given in the following table:

	θ_1	θ_2
d_1	0	100
d_2	100	0

With $p(\theta_1) = 3/4$, $p(\theta_2) = 1/4$, the expected loss with d_1 is 25, with d_2, 75. The smaller of these is 25 dollars, agreeing with the previous calculation of the expected value. Another way of looking at the analysis is obtained by noting that the loss measures what you lose by not knowing the uncertain event.

Perfect information tells you what the uncertain event is, so it reduces the loss to zero. Hence the gain exactly compensates for the loss.

7.6 MINIMIZATION OF EXPECTED LOSS

In deriving this alternative expression for the expected value of perfect information we have obtained an alternative method of solving a decision problem, equivalent to the principle of maximizing expected utility, namely the principle of minimizing expected loss (equation (7.4)). To apply this principle we have to consider, for each uncertain event, what is the best decision, and then to evaluate how much worse the other decisions are compared with that optimum one, the comparison being measured by the loss. The advantages of this procedure are

(1) It only requires a comparison of consequences in each column of the decision table, and not an overall comparison of consequences;
(2) The numbers are usually smaller and the arithmetic consequently simpler (the simple investment example just considered provides an illustration); and
(3) It provides directly an expression for the expected value of perfect information.

In certain applications of decision theory, for example in studying some statistical problems, it is usual to work entirely in terms of losses, and the statistician typically minimizes expected loss.

To counter these advantages there is one difficulty that weighs against the loss concept and which effectively reduces its use to a special case. It has repeatedly been emphasized that utility is not just any number measuring the desirability of a consequence, it is a measure on a probability scale. Now loss, as defined in equation (7.3), is a difference of two utilities: the largest utility possible for the uncertain event minus the utility associated with that event and decision. Consequently loss is a measure of undesirability measured in a rather special way. It is easy to forget this and fail to measure it in a way that makes minimization of expected loss a coherent method of decision-making. The point is best explained by a simple example.

We take the case of our decreasingly risk-averse decision-maker with a scale factor of 10 contemplating the gamble which will win or lose him 100 dollars when his total capital is 100 dollars. (The example was considered in section 5.7.) In dollars the decision table is as follows:

	θ_1: Stock appreciates	θ_2: Stock depreciates
d_1: Invest	200	0
d_2: Leave in bank	100	100

(See Table 5.1.) It is then tempting to argue that the losses are as given in the next table:

	θ_1	θ_2
d_1	0	100
d_2	100	0

reducing the situation to that just considered below equation (7.4) for a decision-maker with 5000 dollars capital contemplating the same gamble.

The argument is false because losses, as properly defined, are differences between utilities, not between dollar amounts. To obtain the loss table for our decision-maker with 100 dollars capital we must first pass to the decision table with utilities instead of dollars. This was found in section 5.7 and is

	θ_1	θ_2
d_1	0.364	0.000
d_2	0.221	0.221

The table with losses is easily seen to be

	θ_1	θ_2
d_1	0.000	0.221
d_2	0.143	0.000

The expected loss for a chance p of the stock appreciating (θ_1) is, for d_1, $0.221 \times (1 - p)$; for d_2, $0.143 \times p$. These are equal when the odds $p/(1 - p)$ are equal to $0.221/0.143 = 1.55$, or about 3–2 on. The corresponding value of p is 0.61, the same value as obtained by the earlier argument. The point is that if the stock appreciates he will drop 100 dollars in selecting d_2 instead of d_1, the drop being from 200 to 100: but if the stock depreciates although he will similarly drop 100 dollars (in selecting d_1 instead of d_2) the drop will be from 100 to zero dollars. For any risk-averse decision-maker the latter drop is more serious than the former and this is reflected in the loss (properly defined) for the stock depreciating (0.221) being greater than in the contrary case (0.143).

7.7 PERFECT INFORMATION WITH ANY MONETARY UTILITY

It is therefore important to remember when using losses that they are differences in utilities and not absolute quantities. With the technical meaning of loss it is not sensible to speak of a loss of 100 dollars, for the loss associated with a fall of 100 dollars in capital depends on the size of that capital. There is one exception to this, namely the case considered at the beginning of this chapter, where the range of monetary values being considered is such that utility is effectively linear in money over that range. Then losses are proportional to differences in dollars and, since a scale factor has no effect on the arguments, such dollar differences may legitimately be used. This is why the specialization was made at first.

Let us next consider the expected value of perfect information for a problem in which the outcomes are entirely monetary, but where the utility of money is not linear. Specifically, take the investment example with capital C and an investment which will either increase or diminish this by an amount a. In utilities the decision table reads as follows:

	θ_1	θ_2
d_1	$u(C+a)$	$u(C-a)$
d_2	$u(C)$	$u(C)$
pr	p	$1-p$

where, as before, $u(C)$ denotes the utility of C, and the other utilities similarly.

Now suppose that a fee f is paid for perfect information. Again we imagine the infallible stockbroker who charges us this amount and in return tells us whether or not the stock will appreciate. The fee is paid irrespective of the outcome and therefore every consequence will have its dollar value reduced by f. Hence the decision table will now be

	θ_1	θ_2
d_1	$u(C+a-f)$	$u(C-a-f)$
d_2	$u(C-f)$	$u(C-f)$
pr	p	$1-p$

With perfect information the expected utility is

$$pu(C+a-f)+(1-p)u(C-f) \tag{7.5}$$

Suppose that without such information d_2 is the better decision, with expected utility $u(C)$. Then the largest fee that one would be willing to pay would be the value of f that makes formula (7.5) equal to $u(C)$. For smaller f formula (7.5) will exceed $u(C)$ and the advice can be expected to be profitable. Consequently the expected monetary value of perfect information is that value f which satisfies

$$pu(C+a-f)+(1-p)u(C-f)=u(C) \tag{7.6}$$

Had d_1 been the better decision without information a similar equation would have been obtained by equating expression (7.5) to the expected utility of d_1, namely

$$pu(C+a)+(1-p)u(C-a)$$

Equation (7.6) may be solved by several methods. For given numerical values for C, a, and p, plus a table or graph of the utility function, the least sophisticated way is simply trial and error with different values of f.

We illustrate again with the decreasingly risk-averse decision-maker contemplating a gamble on 100 dollars with 100 dollars capital. Here $C=100$, $a=100$, and we take the case $p=1/2$ for which the optimum decision is d_2. Equation (7.6) then gives

$$u(200-f)/2+u(100-f)/2=u(100) \tag{7.7}$$

The situation is illustrated in Figure 7.1. In order to emphasize the nature of the argument the scales have been omitted and the curvature of the utility function exaggerated. On the money scale the three relevant values, $100-f$, 100 and $200-f$ are shown. The corresponding utilities are given by the points A, B, and C, respectively, on the utility scale. Equation (7.7) says that B must be the mid-point of AC. (Here B is fixed but A and C can be changed by altering f.) Due to the diminishing marginal utility, $200-f$ will have to exceed 100 by more than 100 will need to exceed $100-f$. It immediately follows that $f=50$ dollars would be too high a fee to pay; since 150 is as much above 100 as 100 is above 50. Consequently values less than 50 should be tried. A trial of a few numbers shows that f around 44 dollars does the trick. $200-f$ is then 156 with utility 0.308 (remembering the scale factor of 10) and $100-f$ at 56 has utility 0.136: their average is 0.222 about equal to that of $u(100)=0.221$. Hence it would be worth our investor paying up to 44 dollars for the advice. The richer decision-maker considered earlier in the chapter with the same problem, except that his capital is 5000 dollars, could go up to 50 dollars. He can afford to pay more for the advice.

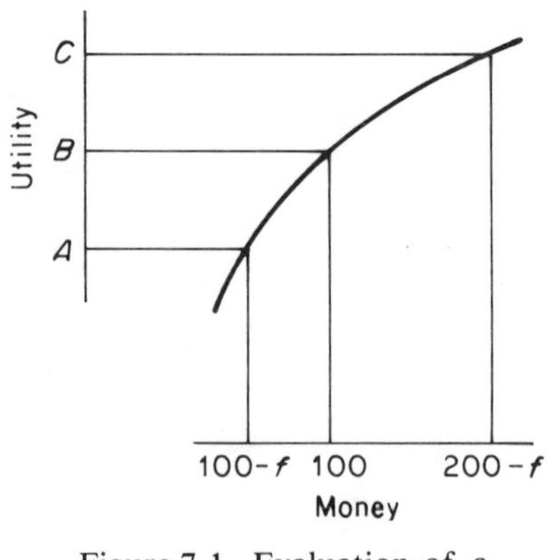

Figure 7.1. Evaluation of a reasonable fee

7.8 GENERAL METHOD FOR PERFECT INFORMATION

The general procedure for situations in which the consequences are entirely monetary should now be clear. The situation without information provides utilities $u(C_{ij})$ and the best that can be done is to maximize expected utility in the usual way. If the perfect information costs, in monetary terms, f, then with the information all the utilities will become $u(C_{ij}-f)$ and the expected utility is given by

$$\sum_{j=1}^{n} \max_i u(C_{ij}-f)p_j$$

(compare equation (7.2)). The monetary value of the perfect information is the solution of the equation which results from equating this value to

$$\max_i \sum_{j=1}^{n} u(C_{ij})p_j$$

the expected utility without information.

In some situations the fee may only be payable with some consequences and not others. Or the fee may vary with the consequences, such as happens when it takes the form of a percentage commission. In these cases the principle is still to modify the consequences C_{ij} to new values $C_{ij}-f_{ij}$ and to proceed in the same way.

If the consequences are not monetary then the effect of the fee has to be combined with the non-monetary elements in an appropriate way to derive new consequences C'_{ij}, say. The fee is worth paying if

$$\sum_{j=1}^{n} \max_i u(C'_{ij})p_j$$

the expected utility that results from the perfect information, exceeds

$$\max_i \sum_{j=1}^{n} u(C_{ij})p_j$$

the expected utility without this knowledge.

7.9 EXPECTED VALUE OF PARTIAL INFORMATION

We now pass from the case of perfect information, where the knowledge changes the probabilities to 0 or 1, to the more general situation where the information is less than perfect and the probabilities are modified by Bayes' theorem in the manner described in the previous chapter. That is, we consider cases where our information affects our beliefs but still leaves some uncertainty present. In particular, we consider the fallible stockbroker whose advice is helpful but not always correct. The argument is a slight extension of that used with perfect information. There we considered in turn each form the perfect information could take, found what was best for each, and then averaged over the possible forms with their appropriate probabilities. The same method works with partial information and it is necessary to contemplate every form that this less than total knowledge could take. As before, it will be advantageous to consider first the situation where the consequences are entirely monetary and the utility of money is linear.

Let us, as in the previous chapter, denote by X the additional data; for example, the actual advice provided by the broker. To be of use, X must be related to the uncertain events θ_j and we suppose the probabilities $p(X \mid \theta_j)$ known: that is, we know the chance of obtaining the data given that θ_j is the true event. In section 6.2 this was called the likelihood function. Consider how the decision-maker should act in the light of X. The effect on his opinions was discussed at length in the last chapter, where we showed that his original probabilities, $p(\theta_j)$, for the uncertain events would be modified to $p(\theta_j \mid X)$, where the latter probabilities are given by Bayes' formula (equation (6.1))

$$p(\theta_j \mid X) = p(X \mid \theta_j)p(\theta_j)/p(X) \tag{7.8}$$

This formula modifies the prior probabilities (prior to X) to posterior probabilities (posterior to X) by multiplying by the likelihood and then scaling by $p(X)$. We cannot, as in section 6.2, use the proportionality form of Bayes' theorem, omitting $p(X)$, for it plays an important role as we now see. Essentially, X is the data, uncertain to us before the advice is requested, and therefore described probabilistically. In section 6.2 it was known and its probability was irrelevant.

The decision-maker has the new probabilities for the uncertain events and his best procedure is to maximize expected utility, the expectation being taken with the posterior values. He will therefore obtain

$$\max_i \sum_{j=1}^{n} u_{ij} p(\theta_j \mid X) \tag{7.9}$$

instead of

$$\max_i \sum_{j=1}^{n} u_{ij} p(\theta_j) \tag{7.10}$$

that he would have obtained without X.

Exactly as in the case of perfect information, where the decision-maker did not know what form the information would take, here he does not know what data will arise. Nevertheless he does have probabilites for the data. He knows the probabilities for the uncertain events $p(\theta_j)$ and he knows the probability of X for each event, $p(X \mid \theta_j)$. Consequently, extending the conversation from X to the θ_j

$$p(X) = \sum_{j=1}^{n} p(X \mid \theta_j)p(\theta_j) \tag{7.11}$$

This probability is exactly that occurring in Bayes' formula (equation (7.8)). (The reader should be careful to distinguish between $p(X \mid \theta_j)$, the probability of X for a given uncertain event θ_j, and $p(X)$, the overall probability of X given the original decision-making situation.)

The decision-maker knows what to do if he obtains X, namely to use formula (7.9), and he knows the probability of X; consequently what he can expect to obtain is obtained by multiplying formula (7.9) by $p(X)$ and summing over the various possible forms the data X can take, in the usual manner for calculating an expectation. The result is

$$\sum_X \max_i \sum_{j=1}^{n} u_{ij} p(\theta_j \mid X) p(X)$$

But from Bayes' result (equation (7.8))

$$p(\theta_j \mid X) p(X) = p(X \mid \theta_j) p(\theta_j)$$

so that the expected utility which will result from the partial information is

$$\sum_X \max_i \sum_{j=1}^{n} u_{ij} p(X \mid \theta_j) p(\theta_j) \qquad (7.12)$$

The original value was

$$\max_i \sum_{j=1}^{n} u_{ij} p(\theta_j) \qquad (7.13)$$

and the difference between these is the *expected value of partial information*. The formulae are easy to apply, since the only quantity occurring in formula (7.12), not already present in the original problem described in formula (7.13), is the likelihood function $p(X \mid \theta_j)$. We now illustrate the use of these results.

7.10 INVESTMENT EXAMPLE REVISITED

Consider the original example at the beginning of this chapter of someone with a capital of 5000 dollars contemplating a gamble with equal chances of winning or losing 100 dollars and with a utility proportional to money. We saw that perfect information was worth 50 dollars. Suppose now that he contemplates seeking the advice of a fallible broker whose advice cannot always be relied on but who nevertheless is better informed than the decision-maker. How much is such advice worth? Obviously less than 50 dollars: but how much less? It clearly depends on the degree of fallibility, some brokers being better than others, and before we can make any progress we have to describe the broker's ability quantitatively. How is this to be done? The answer is simple: we have already seen that the additional information needed in the general argument just described is contained in the likelihood function, $p(X \mid \theta_j)$, and it is only necessary to note that $p(X \mid \theta_j)$ is just such a quantitative description of the broker's reliability. To see this, notice that the broker can either advise purchase (on the grounds that θ_1 obtains and the stock will appreciate) or not (thinking that it will depreciate); so that there are two possible pieces of data, two possible values of X, namely 'purchase' or 'do not purchase'. Denote these by X_1 and X_2, respectively. Then $p(X_1 \mid \theta_1)$, for example, is the probability of the broker giving the advice to purchase when the stock will rise; in other words, correct advice. Similarly, $p(X_2 \mid \theta_2)$ corresponds to another piece of good advice, whereas the complementary probabilities $p(X_1 \mid \theta_2)$ and $p(X_2 \mid \theta_1)$ are the probabilities of incorrect recommendations. We saw in section 6.9 how these errors of the two kinds described the behaviour of the broker and how his advice caused the probabilities associated with the uncertain events to be modified.

Consider, then, a fairly reliable broker who gives the correct advice 3/4 of the time: specifically suppose

$$p(X_1 \mid \theta_1) = p(X_2 \mid \theta_2) = 3/4$$

and hence

$$p(X_2 \mid \theta_1) = p(X_1 \mid \theta_2) = 1/4$$

The necessary calculations are shown in Table 7.2. The upper left-hand part of the table is the usual decision table. Underneath the prior probabilities are entered the likelihoods, one row for X_1, another for X_2. For any decision d_i and result X we calculate part of formula (7.12), namely

$$\sum_{j=1}^{n} u_{ij} p(X \mid \theta_j) p(\theta_j)$$

obtained by taking each column (that is, each value of j) multiplying the utility for d_i, likelihood for X, and prior probability together, and adding the results. Thus for d_1 and X_1 we have from the table

$$5100 \times 3/4 \times 1/2 + 4900 \times 1/4 \times 1/2 = 2525$$

the two terms on the left coming from the two columns, θ_1 and θ_2, of the table. This value is entered on the right of the table in a column headed X_1 and in the row corresponding to d_1. The other values for each decision and each X are similarly entered. For each X, that is, in each right-hand column, the largest is selected. In the table these are indicated by underlining. Finally the results for the different X's are added together. The value for Table 7.2 is 5025

Table 7.2

	θ_1	θ_2	X_1	X_2
d_1	5100	4900	$\underline{2525}$	2475
d_2	5000	5000	2500	$\underline{2500}$
$p(\theta_j)$	1/2	1/2		
$p(X_1 \mid \theta_j)$	3/4	1/4		
$p(X_2 \mid \theta_j)$	1/4	3/4		

Table 7.3

	θ_1	θ_2	X_1	X_2
d_1	0	100	12.5	37.5
d_2	100	0	37.5	12.5
$p(\theta_j)$	1/2	1/2		
$p(X_1 \mid \theta_j)$	3/4	1/4		
$p(X_2 \mid \theta_j)$	1/4	3/4		

dollars; this is the numerical value of expression (7.12). Without the advice either decision produces an expectation of 5000 dollars (formula (7.13)), consequently one would be prepared to pay up to the difference, namely 25 dollars for the advice. This is the expected value of the broker's advice. Notice that this is only one-half the value of perfect information: our fallible broker has, in this problem, only half the value of the ideal advisor.

The calculations may also be carried through in terms of losses, which in this case, with linear utility for money, are just differences of monetary gains. The details are given in Table 7.3. Notice that the numbers are smaller than those in the earlier table. With losses we are concerned with minimization, not maximization, so that in each column headed X the least value will be required. In Table 7.3 the sum of these minima is 25 dollars. The expected loss without advice is 50 dollars, so again the difference is 25 dollars. We postpone until section 7.16 a proof that, as with perfect information, it is legitimate to calculate with losses.

7.11 INFORMATION IS ALWAYS EXPECTED TO BE OF VALUE

At the end of this chapter there is also a proof that the expected value of partial information is greater than zero. (It can be zero but only if the information is irrelevant.) The result is of some practical importance, so let us be clear just what is being claimed. If you are about to receive some information relevant to some uncertain events, then the expected utility for *any* decision problem involving those events is expected to increase as a result of the information. Notice there are two expectations in that sentence: the first is concerned with the uncertain events, the second with the uncertain information. More briefly, the information is always expected to be of value. Of course, if you have to pay for it then the payment may bring its value down below zero: thus in the first example of section 7.2 with expected value of 50, a fee of 75 dollars would make the expectation a loss of 25 dollars. So we should say: cost-free information is always expected to be of value. Now let us see some implications of this.

In section 3.11 we saw how the law should use information, or what it calls 'evidence'. For the decision problem of the defendant's guilt our result says that any relevant evidence is expected to be of value. Now the law has certain rules for the admissibility of evidence, one criterion being the cost of the evidence. Our result says that the cost should be the *only* reason for ruling evidence to be inadmissible: that if evidence is virtually cost-free then it should be admitted, for it is expected to be of value in judging the case. This goes against current legal practice. Thus English law does not allow evidence of bad character to be used to increase the probability that the defendant is guilty. Our argument says it should if the evidence is almost free. We are not saying that the law is incorrect: we are saying that the situation needs reconsideration in the light of expected utility theory. It may be, for reasons that are not clear to me, that some material fact has been forgotten in applying the theory to legal practice.

Another interesting application of the result is to the behaviour of governments and the freedom of citizens to have access to information about their actions. It is probably cheaper to publish information rather than attempt to ensure its secrecy by employing security devices. So the provision of information is virtually cost-free. Consequently governments should be open with their people and not deprive them of information. It is absurd that a citizen should not have completely free access to information held about him on any computer; social service, police, or credit company. It is almost equally absurd that he should not have information about his neighbour, though here two new points arise.

One reason for denying access to information is that people do not know how to use it. This argument has been used in law. It has some merits but they disappear within the expected utility thesis because that thesis contains an account of how to use the information, namely by use of the rules of probability.

A second reason for denying access to information is correct, even within our thesis. This argument uses the concept of two decision-makers, for example, a government and the government of a hostile country. To have open information would be prejudicial to the security of the state. Similarly for me to have information about my neighbour might give me an advantage over him. These ideas merit attention because the maximum expected utility thesis is limited to a single decision-maker. It does not refer to two hostile armies. This is a major point, and we return to discuss it in Chapter 10.

7.12 COMPARISON BETWEEN DIFFERENT INFORMATION SOURCES

These ideas enable one to attach an expected monetary value to any information, at least in the case where the consequences are entirely monetary and the utility is linear therein. If the information is available at a price less than this, it is best to take it, otherwise it should be refused. In this way it is possible for an engineer to put a value on experiments, costly in apparatus and manpower, intended to help him design a piece of equipment. Only if the investment in the experiment is below the expected monetary return will it be

Table 7.4

	θ_1	θ_2	X_1	X_2
d_1	0	100	20	30
d_2	100	0	45	5
$p(\theta_j)$	1/2	1/2		
$p(X_1 \mid \theta_j)$	0.9	0.4		
$p(X_2 \mid \theta_j)$	0.1	0.6		

worthwhile. Similarly, a firm contemplating the marketing of a new product can determine whether it is worth accepting the estimate of a market research firm for carrying out a survey designed to estimate the likely demand for the product. A further use for these ideas lies in the comparison of the likely effectiveness of two different types of information. Thus the engineer might contemplate carrying out some laboratory experiments or, alternatively, building a prototype. If the costs of these are comparable then he could decide between them on the basis of the expected return from the data they might produce. We illustrate this latter type of argument, using again the technically simpler investment example.

The broker in section 7.10 was reliable to the extent of giving correct advice three times out of four, irrespective of whether the stock appreciated or not. Now consider a second broker who is excellent at spotting a good stock but poor at detecting a loser. Remembering that X_1 is the advice to buy and θ_1 the event that the stock improves, suppose that for the new broker $p(X_1 \mid \theta_1) = 0.9$ but that $p(X_2 \mid \theta_2) = 0.6$. We saw that a reasonable fee for the first broker when consulted about our investment problem was 25 dollars. How does this compare with the other broker; is his advice worth more or less? The calculations are shown in Table 7.4. The total of the underlined values for formula (7.12) is 25 dollars, so the advice is again worth 25 dollars. Despite differences in their behaviour the brokers are of equal merit. Of course, with other investment problems the comparison between these two men could easily be different. The reader is invited to try other examples.

7.13 INFORMATION WITH ANY MONETARY UTILITY

If the utility of money is not linear but the consequences are still monetary, then it is necessary to proceed as with perfect information and deduct the fee f from each consequence before evaluating the utility. Expression (7.12) will then become

$$\sum_X \max_i \sum_{j=1}^{n} u(C_{ij} - f) p(X \mid \theta_j) p(\theta_j)$$

assuming a constant fee. The monetary gain to be expected from the information is the value f which makes this expression equal to the original expected utility given by formula (7.13). The calculations here are somewhat more involved because of the complexity of the resulting equations, and are therefore omitted, but the principle remains the same.

The situation where perfect information can be acquired is rare, whereas partial information is often obtainable. Nevertheless, the expected value of perfect information is useful because it provides an upper bound to that for partial information. In other words, the former is never less than the latter. For example, suppose someone offers you some partial information for 100 dollars. If the expected value of perfect information is only 80 dollars then the offer is not worth accepting. The calculation for perfect information is straightforward, whereas that for partial information is more complicated and additionally involves assessment of the reliability of partial knowledge: in technical terms, assessment of the likelihood functions. Hence the simpler analysis in this case rules out the need for the more complicated. Of course, had the expected value of perfect information been 200 dollars, say, then more consideration would have had to have been given to the offer.

The proof of the result stated in the last paragraph is a straightforward extension of the arguments used at the end of this chapter and is left as an exercise for the interested reader.

7.14 CHOICE OF HOW MUCH SAMPLING TO DO

In section 6.10 we discussed the case of the manufacturer sampling a batch in order to gain partial information about its quality. This is an example where the expected monetary value of the sample, expressed through the expected value of partial information, could be assessed and compared with the actual cost of the sampling. Furthermore, it would be possible to determine the best size of sample to take. It is typically true that as the sample size increases the expected value of the partial information, which naturally increases, does so with diminishing marginal value. For example, the value to be expected from increasing the sample size from 5 to 6 will be greater than that resulting from increasing it from 6 to 7. Information (usually but not always) behaves like utility in exhibiting decreasing marginal values with sample size. Consequently

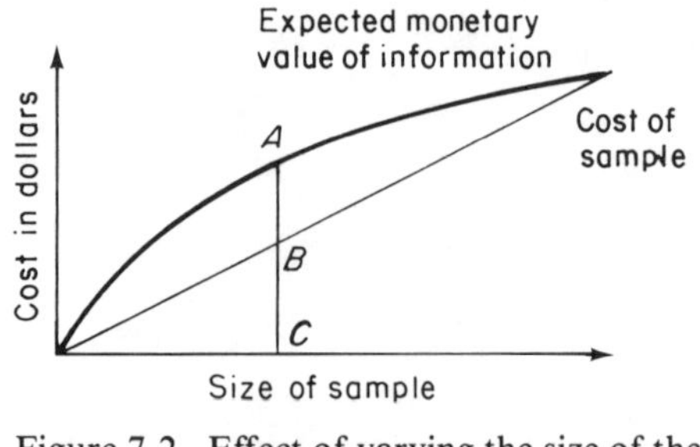

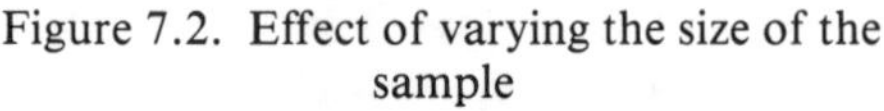
Figure 7.2. Effect of varying the size of the sample

there will come a stage when the marginal increase is not worth the additional cost of sampling. The situation is shown schematically in Figure 7.2. The straight line shows a typical relationship between the size of sample and its cost, where each item costs the same amount to inspect. The curve shows the usual form for the expected monetary value of the information from a sample, where it increases with sample size but its rate of increase diminishes. The point C provides the optimum sample size since there the distance AB between the curve and the straight line is a maximum. Beyond C, the increase in information (shown by A) is not enough to overcome the cost of the sample (described by B). Again we do not enter into the calculations because they are technically tedious, but hope that enough has been said to explain the principles upon which they are based. There is no difficulty, for example, in writing a computer program to determine the best value.

7.15 SEQUENTIAL INFORMATION

In this example we have considered the manufacturer taking a sample of fixed size (in the previous chapter it was 10) and then reaching a decision, and have then gone on to consider the optimum size. There is an alternative way of proceeding which is usually better. In this method the samples are taken from the batch one at a time, inspected, and then, after each inspection, a decision is made about whether to continue sampling or to conclude that enough is known about the batch and further sampling inspection is unnecessary. This is in distinction to the situation so far contemplated, where the decision is to be delayed until the whole of the sample has been seen. It is easy to see that the new method is likely to be better by considering what happens if, in taking a sample of size 10, the first 6 items, say, are all defective. This is probably enough warning that the whole batch is of poor quality and therefore that the further 4 items are hardly worth looking at. In other words, in some cases it may be possible to reach a decision before all the results are in. This new method is called *sequential*, because the items are inspected in sequence and a decision made after each one.

Many decision problems are sequential in character. The engineer designing and building a plant would not have to make one overall decision and then adhere to it. He could probably make a few initial decisions, leaving others until later in the design and construction process, when he has more information. The whole operation would naturally consist of a sequence of interconnected decision problems. In the first chapter of this book we mentioned an example of a retailer having to decide how many items to order for his stock. Typically this decision will be related to the order he will have to place the following week and to the order he placed last week. He really has a weekly sequence of decision problems, all interrelated. The rules of coherent behaviour that have been developed in this book will continue to apply to decisions conducted in sequence and lead to optimum rules of procedure. The principal tools in handling such problems are the concept of a decision tree and the consequent ideas of dynamic programming. These form the major topics of the next chapter.

7.16 TWO PROOFS

The present chapter concludes by proving two results omitted in the main text. The proofs may be ignored by readers prepared to take them on trust. The first says that information, even when incomplete, can be expected to have positive utility: the second says that losses may be used in evaluating the expectation. Both proofs are generalizations of those used in deriving the similar results for perfect information.

With partial information the expected utility is given by formula (7.12): write the sum over j as $u_i(X)$ so that formula (7.12) is simply

$$\sum_X \max_i u_i(X)$$

This is certainly not less than

$$\sum_X u_i(X)$$

whatever value of i is selected in this last expression, and therefore in particular it is not less than

$$\max_i \sum_X u_i(X)$$

But now putting back the full expression for $u_i(X)$, this last formula becomes

$$\max_i \sum_X \sum_{j=1}^{n} u_{ij} p(X \mid \theta_j) p(\theta_j)$$

However,

$$\sum_X p(X \mid \theta_j) = 1$$

since the X's are exclusive and exhaustive. Consequently the final expression is

$$\max_i \sum_{j=1}^{n} u_{ij} p(\theta_j)$$

which is exactly formula (7.13), the expected utility without information. That is, formula (7.12) is not less than formula (7.13), which is what we require.

The two expressions will be equal, and therefore the difference zero, only if

$$\sum_X \max_i u_i(X) = \max_i \sum_X u_i(X)$$

for the point where we said the former was not less than the latter is the only point in the argument where a difference could occur. Suppose the decision that maximizes the right-hand side is d_s, so that

$$\sum_X \max_i u_i(X) = \sum_X u_s(X)$$

This can only happen if

$$u_s(X) = \max_i u_i(X)$$

for all X: that is, if the optimum decision does not depend on X. In other words, the only case where information is not expected to be worth having is where it is going to be ignored, whatever it is.

To see that losses may be used, subtract from u_{ij} any value a_j: that is, any value that depends on the event but not the decision. Then $u_i(X)$ is reduced by

$$\sum_{j=1}^{n} a_j p(X \mid \theta_j) p(\theta_j)$$

which does not depend on i, and so is unaffected by the maximization over i. Then summation over X, for the same reason as before, gives

$$\sum_{j=1}^{n} a_j p(\theta_j)$$

as the amount by which formula (7.12) is reduced. But formula (7.13) is also affected by the same quantity, and hence their difference, which is the expected value of partial information, is unaffected. It only remains to put

$$a_j = \max_i u_{ij}$$

to obtain the required result, since the losses are, by definition (equation (7.3)),

$$l_{ij} = \max_i u_{ij} - u_{ij}$$

Exercises

7.1. Calculate the expected value of perfect information in the following situations: (the entries are dollars and utility may be supposed linear in money—it will be simplest to work in terms of losses).

(i)

	θ_1	θ_2
d_1	130	90
d_2	100	105
pr	1/3	2/3

(ii)

	θ_1	θ_2
d_1	1000	100
d_2	100	200
pr	0.2	0.8

(iii)

	θ_1	θ_2	θ_3
d_1	500	300	200
d_2	400	200	400
pr	0.2	0.5	0.3

7.2. Recalculate the expected value of perfect information in Exercise 7.1(ii) for the decreasingly risk-averse decision-maker with a scale factor of 10.
7.3. A firm is considering launching one of two products, d_1 or d_2. The former is expected to do well only if the economy is good, the latter may do better if the economy is weak. The estimated returns (in 1000 dollars) and the relevant probabilities are

	θ_1: good	θ_2: moderate	θ_3: weak
d_1	50	30	20
d_2	40	30	30
pr	0.3	0.4	0.3

A market research firm offers to provide for 4000 dollars a survey of the prospects. Should the offer be accepted? (Utility may be supposed proportional to money.)
7.4. Consider a perfect stockbroker who only charges a fee if stock is purchased (d_1). Obtain an equation analogous to equation (7.6) for his fee and show that a reasonable fee is equal to the amount by which the stock might appreciate. This fee would mean that you always ended up with C, whatever happened, and whatever decision you took.
7.5. (i) For Exercise 7.1(i) calculate the expected value of advice which is correct with probability 2/3.
(ii) For Exercise 7.1(ii) calculate the expected value of advice which is always correct if θ_2 is true, but is only 50% correct if θ_1 is.
(iii) For Exercise 7.1(iii) calculate the expected value of advice which gives X_1, X_2, X_3 with

$$p(X_i \mid \theta_i) = 1/2, \quad p(X_i \mid \theta_j) = 1/4, \; i \neq j$$

(In words, the advice is equally likely to be right or wrong, and when wrong both mistakes are equally likely.)
7.6. Calculate the expected value of the advice suggested in Exercise 7.5(iii) when applied to Exercise 7.3.

Chapter 8

Decision Trees

'To decide is useless if decision cannot be carried out.'

Flowering Wilderness, Ch. 27.

8.1 INTRODUCTION

The decision problems so far studied have been described in terms of decision tables and the analysis has utilized this tabulation. The tables have the structure that the rows correspond to decisions, the columns to uncertain events, and the entries in the body of the tables to the consequences and their utilities, the probabilities being conveniently placed in an additional row at the foot of the table. The analysis proceeds by calculating for each row (decision) the expected utility and selecting that act with maximum value. Essentially the analysis is based on breaking the whole table down into separate small decision problems in which each consequence is replaced by a gamble at the best and worst consequences. Once the separate decisions of how a consequence can be replaced by a gamble have been made, the remainder of the calculations only consist in fitting the pieces together. Indeed, our whole concept of coherence is nothing more than this fitting together of separate decision problems.

Now we have seen in discussing the utility of money that other ways are open to us in describing the consequences: they do not always have to be related to the best and worst outcomes. Our aim in this present chapter is to describe another way in which a decision problem can be broken down into smaller problems which can be solved separately and then combined to provide a solution to the larger problem. The device is especially useful when the grand problem can be broken down into a sequence of problems which follow one another in some natural order. In section 7.15 it was explained how such situations often occur. The method of analysis uses a decision tree.

Like a real tree, a decision tree contains several parts that act together, or cohere. Our method solves the easier problems that occur at the different parts of the tree and then uses the rules of probability to make them cohere. As before, it is the coherence that is the principal novelty and the major tool in what we do. We begin by describing a decision tree for a modified form of the simple investment example in section 7.2.

8.2 THE INVESTMENT EXAMPLE REVISITED

The particular example has the same structure as before and for convenience the tabular form is repeated:

	θ_1: Stock appreciates	θ_2: Stock depreciates
d_1: Invest	5100	4900
d_2: Leave in bank	5000	5000
pr	0.6	0.4

The investor has 5000 dollars capital and stands to win or lose 100 dollars by the venture. To avoid having too many symmetries in the problem and thereby giving it a special flavour, the case where the probability of appreciation is 0.6 (and not 0.5) will be taken. The utility of money is supposed linear over the range of dollars involved. The investor has available the services of a broker who for a fee of f dollars (f is left unstated for the moment) will advise. His advice will either be to buy the stock, X_1, or not, X_2. The broker's reliability is described by the following probabilities,

$$p(X_1 \mid \theta_1) = 0.8, \quad p(X_2 \mid \theta_2) = 0.7$$

In other words, our investor's assessment of the broker's ability is that he is a little better at spotting a winner than a loser, having 80% chance of being right with the former but only 70% with the latter. The investor has to decide whether or not to buy the advice and whether or not to invest (either with or without advice). This could be presented as in section 7.2, or as a table with many decisions, of which a typical one is 'pay the fee for the advice and if it is to buy, accept the advice'. Or finally, as we shall now demonstrate, as a decision tree.

The decision tree for this problem is given in Figure 8.1. All our trees grow horizontally from left to right; the trunk is at the left of the page and the branches to the right. Beginning at the left, the investor has three acts available to him. These are described by three branches of the tree; the first, labelled broker, means paying for his advice; the second corresponds to deciding to invest without advice, d_1, and the third to leaving the money in the bank, d_2. (In Figure 8.1 the reader should ignore for the moment all the numbers and all the slashes through the branches.) Let us pursue the branch labelled broker. If advice is sought, two things can happen: either the advice can be to buy, X_1, or not, X_2. Consequently this branch now breaks into two, labelled X_1 and X_2, respectively. Whichever one of these last two branches is followed, the investor has finally to select between d_1 and d_2, so each of these eventually breaks into two further branches so labelled. This completely describes the decision structure of the problem, but at the end of each of the present terminal branches (each of which is either d_1 or d_2) it is necessary to add two others,

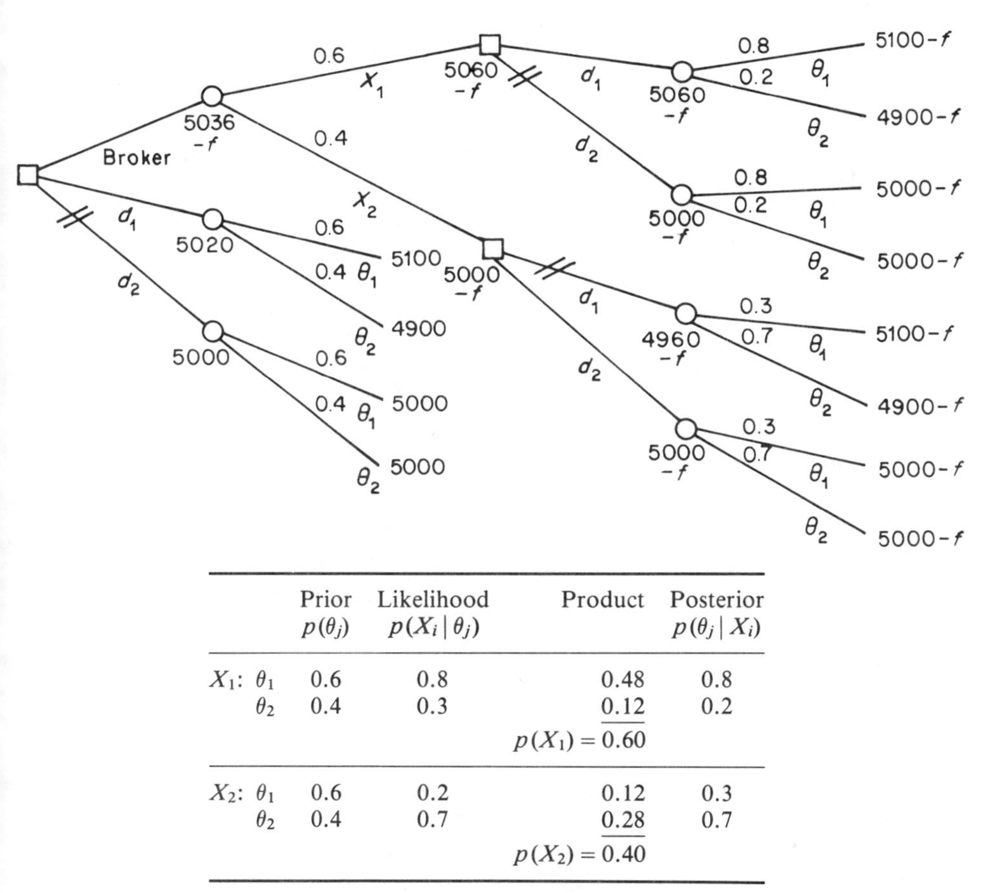

	Prior $p(\theta_j)$	Likelihood $p(X_i \mid \theta_j)$	Product	Posterior $p(\theta_j \mid X_i)$
X_1: θ_1	0.6	0.8	0.48	0.8
θ_2	0.4	0.3	0.12	0.2
			$p(X_1) = 0.60$	
X_2: θ_1	0.6	0.2	0.12	0.3
θ_2	0.4	0.7	0.28	0.7
			$p(X_2) = 0.40$	

Figure 8.1. Decision tree for the investment problem

one labelled θ_1, the other θ_2, corresponding to whether the stock appreciates or not. We now have the complete tree consisting of a series of branches, each branch described by either a decision or an observed result. Notice that in describing the tree from left to right the natural order of the events in time has been followed, so that at any point of the tree the past lies to our left and can be studied by pursuing the branches down to the trunk, and the future lies to our right along the branches springing from that point and leading to the tips of the tree.

8.3 DECISION AND RANDOM NODES

Consider next the points where the branches split into other branches: these are called *nodes*. The nodes are of two distinct types that we now describe. Beginning again at the left of the tree, the first node leads to three branches, 'broker', d_1 and d_2, and the choice of which branch to follow depends on the decision-maker who can select from the three possibilities. Consequently this node is called a *decision node*, and is represented by a square. Following any one of these three branches, say that labelled 'broker', we reach a second node from which two branches, X_1 and X_2, emanate. The decision-maker at this node has no control over which branch is selected (in this case the broker makes the choice) and the behaviour is therefore different from that at a decision node. We have seen in the previous chapter that the investor can attach probabilities to the form of the advice, so this node is called a *random node*, and is represented in the tree by a circle. Following either of the branches X_1 or X_2 we come to decision nodes, where the investor selects d_1 or d_2. Finally, each of the branches labelled d_1 or d_2 ends in a random node, where the investor has no control over whether the stock appreciates or not. (It is attractive, but not particularly useful, to say that 'nature' makes the selection between θ_1 and θ_2.)

A typical decision tree therefore consists of a series of branches stemming from nodes of two types, decision or random. Moving along the branches from left to right the two types of node alternate, beginning with a decision node and ending with a random one. Each branch is labelled: all those springing from a decision node, with a decision; the others from random nodes being described by uncertain events. Such trees can be written down for any decision problem, being particularly useful when the problem, as here, breaks down into a sequence of separate problems. We now describe the method of analysing a tree.

8.4 PROBABILITIES AT RANDOM NODES

The analysis will involve two types of quantities, probabilities and utilities. We begin with the former by calculating the probabilities associated with the branches emanating from the random nodes. It is important to remember that our tree develops in time order from left to right and therefore at any node we have all the information prior to that position—that is, back from that node to the base of the tree—and none of it that comes after—that is, to the tips of the tree. Also the probabilities at a node are conditional on our knowledge at the node. Begin at the base of the tree and proceed along any branch, for example that labelled d_1. We immediately reach a random node where the branches correspond to θ_1 and θ_2, and since no advice has been sought or obtained the probabilities are just the initial ones given in the decision table above, namely $p(\theta_1) = 0.6$, $p(\theta_2) = 0.4$. The branches from the random node have been labelled accordingly. A similar argument holds for the decision branch corresponding to d_2.

Along the third branch from the base of the tree the argument is different. Here advice is sought and at the first random node the split depends on what that advice is: to buy, X_1, or not, X_2. We require the probabilities associated with these two values. At this point we do not know whether θ_1 or θ_2 obtains so that the probabilities $p(X_1 \mid \theta_1)$ and other similar ones are not what is required. All we have is our initial knowledge which, in accord with the convention explained in section 3.2, is always with us and is therefore omitted from the notation. The probability needed is consequently $p(X_1)$, with $p(X_2) = 1 - p(X_1)$.

This may easily be calculated by extending the conversation to include θ_1 and θ_2, since probabilities conditional on these values are known. We have

$$p(X_1) = p(X_1 \mid \theta_1)p(\theta_1) + p(X_1 \mid \theta_2)p(\theta_2)$$

and the calculations are shown in the table associated with the tree. In the section of the table labelled X_1, the first column lists the prior probabilities, $p(\theta_1)$ and $p(\theta_2)$, and the second provides the likelihoods, $p(X_1 \mid \theta_1)$ and $p(X_1 \mid \theta_2)$, for the advice X_1 under the two possibilities (notice these latter do not typically add to 1). The next column gives the corresponding products of the priors and likelihoods, which are the two terms on the right-hand side of the equation just cited. Their sum is the required $p(X_1)$ and is provided by adding the entries in the third column. A similar calculation may be performed for X_2, though this is strictly unnecessary since $p(X_2) = 1 - p(X_1)$, but the results will be needed in another connection below and so are worth having. (The fourth column of the table may be omitted for the moment.) The arithmetic shows that $p(X_1) = 0.6$, $p(X_2) = 0.4$: that is, the chance of the broker advising purchase of the stock is 60%. These values are inserted on the appropriate branches emanating from the random node.

Continuing along these branches we come to decision nodes where it is necessary to decide, having the appropriate advice. Passing these by we arrive at random nodes corresponding to θ_1 and θ_2, like the random nodes we first considered. Let us take for illustration the uppermost one in the tree, that corresponding to advice to buy (X_1) and a decision to accept that advice (d_1). The relevant probabilities are now $p(\theta_1 \mid X_1)$ and $p(\theta_2 \mid X_1)$, the posterior probabilities of θ_1, and θ_2 given X_1, that is, given all the information in the tree from that random node back to the base. These are easily calculated by Bayes' theorem,

$$p(\theta_i \mid X_1) = p(X_1 \mid \theta_i)p(\theta_i)/p(X_1)$$

for $i = 1$ and 2. The products in the numerators on the right-hand sides are just those that have been found in the third column of our table, and the denominators are their sums, also already found. Consequently most of the arithmetic has already been done and all that is necessary is to divide one by the other. The entries are given in the fourth and final column of the table: for example, in the first row, $0.8 = 0.48/0.60$. Hence $p(\theta_1 \mid X_1) = 0.8$; given the advice, the chance of the stock appreciating has grown from its original value of 60% to 80%, and this because one believes the broker capable of recognizing a good stock. Similar calculations can be made for the other branches and the results entered along the branches springing from the random nodes.

The principle behind the calculations so far made is simple. At every random node the branches starting there have associated with them probabilities of the events corresponding to the branches, conditional on all the knowledge contained in the tree between that node and the base of the tree. The probabilities in the different parts of the tree have to cohere, the coherence being achieved by the laws of probability. This is the general principle behind the probability assessments in any decision tree.

8.5 UTILITIES AT DECISION NODES

Having completed the probability calculations for the tree let us next consider the utilities or, in the special circumstances of this problem, the money values, since utility is here supposed linear in money. The terminal branches of the tree will have associated with them utilities describing the consequences that result from the selected actions and the uncertain events. For example, in the top right-hand corner, advice X_1 has been obtained, d_1 selected, and the stock appreciates with a capital increase to 5100 dollars less the fee, f, charged for the advice. All these utilities can be obtained from the original formulation as a decision table. The utilities at places other than the terminal branches will follow from the calculations now to be described.

8.6 ANALYSIS OF A DECISION TREE

With all the probabilities and utilities entered into the appropriate places in the tree the analysis can now be completed. The principle is to start from the terminal branches and work back to the base, at each random node calculating an expected utility, and at each decision node choosing that branch with maximum expected utility; the two operations of expectation and maximization alternating back to the base. In our tree suppose we start at the top right-hand corner. Here is a random node with chance 0.8 at 5100 dollars and 0.2 at 4900 (less, in either case, the already incurred fee). The expectation is $0.8 \times 5100 + 0.2 \times 4900 = 5060$ dollars. This is entered at the random node. The next random node below it similarly gives an expectation of 5000 dollars, less fee. Moving down the tree from these two random nodes, a decision node is encountered. The choice is between d_1 and d_2; the former will yield, according to the calculations just made, 5060 dollars, the latter, 5000, both less the fee. Maximization decrees that we choose the former: $5060 - f$ can be entered at the node and the decision d_2 can be forgotten—this is indicated by a slash through the branch. The decision node below it similarly yields $5000 - f$ through d_2. Moving back from these nodes we reach another random node: here the expectation is $0.6 \times 5060 + 0.4 \times 5000 = 5036$, less the fee. Similar calculations from the terminal branches not so far considered give, for d_1, 5020, and for d_2, 5000 dollars. At the base of the tree, a decision node, there is a choice between seeking the broker's advice at an expectation of $5036 - f$, investing, at 5020, and leaving it in the bank, at 5000. Clearly the last is out and the first is worth taking if the fee is less than 16 dollars, which is the expected value of the information that the broker can provide. For example, if $f = 8$ the advice should be sought and the whole decision is worth 5028 dollars: if $f = 20$ then it is best to invest without consultation with an expected reward of 5020 dollars. The calculations may alternatively be carried through in terms of losses with a consequent saving in arithmetic.

In summary, the decision tree method proceeds in the following stages:

(1) The tree is written out in chronological order, the decisions and events being described by branches in the order in which they occur;
(2) Probabilities are attached to the branches emanating from random nodes in any coherent and convenient way;
(3) Utilities are attached to the terminal branches;
(4) Proceeding back from the terminals to the base, by taking expectations at random nodes and maximizing at decision nodes, the best decisions and their expected utilities are determined.

The method is of wide applicability and admirably adapted to modern computers, involving only two operations in the calculations in stage (4), namely maximization and expectation. The main programming difficulty lies in storing the information in a conveniently accessible way. In applications stages (1) to (3) are the difficult ones, involving a description of the tree and the quantitative description of beliefs and values.

There is a moral to be learnt from the analysis. Notice that the calculation in the final stage (4) proceeds from terminals to base; that is, from the events occurring last in time, back to the decisions occurring first. In other words, the problems that arise *last* are the *first* to be considered; those that are the *first* to happen are the *last* to be computed. We cannot decide whether to ask our broker for advice until we have decided what to do with the advice when we get it. In the particular tree studied here the advice is accepted—d_1 if X_1, d_2 if X_2—and the expectations calculated accordingly, with the conclusion that the advice should be worth 16 dollars. The point is succinctly put. We cannot decide what to do today until we have decided what to do with the tomorrows that today's decisions might bring. This is an unfortunate, but indisputable, conclusion. Alternatively expressed, our advice is the opposite of that offered by the King of Hearts to the White Rabbit: 'Begin at the end and go on till you come to the beginning: then stop.'

This completes the study of the investment problem and the description of the general procedure for analysing decision trees. In the remainder of this chapter other examples of trees will be studied.

8.7 GOOGOL, OR THE SECRETARY PROBLEM

The next example has numerous applications: we mention just two, describing the problem and its solution in a straightforward context and only later mentioning its more doubtful sociological consideration. The reason for considering the example at all is that it illustrates a common situation where a very complex situation, and therefore a large tree, can be broken down into small subproblems which are essentially alike. The solution of the common subproblem, plus coherent fitting together of the parts, produces the solution to the complex problem. The subproblems are like bricks, the concrete provides the coherence.

The game of 'googol' is played by one player taking a number of pieces of paper, say N of them, and writing on each piece a number. The numbers can be as large or as small as he pleases, though, for convenience, they should all be different. The pieces are placed, face downwards, on a table. A second player turns up the pieces one at a time and inspects the numbers. At any stage the second player may stop and declare the last piece of paper turned up to have the largest of all the N numbers written upon it. Notice that he may only say this of the *last* piece: he cannot judge that an earlier piece was, after all, the largest. Once a piece has been passed by it is out of the game. If he gets to the final piece he must declare that to be the largest. The prizes vary with the version of the game. The simplest is to award a prize only if the second player is correct, but other variants have second or third prizes, and in one version he gets a prize of k dollars if the selected number has rank $N-k$: thus the largest number gives him $N-1$, the next $N-2$, and so on, only the selection of the smallest giving him no reward. In any version the problems are; first, to determine the optimum strategy for the second player, and second, to find out how much player two can expect to win by employing this strategy. Such an expectation might represent a suitable admission fee.

If the prizes depend only on the ranks of the numbers and not on their absolute values, the first player has no strategy problems. We therefore consider only the second player.

The decision tree is vastly more complicated than the earlier one, but only because it contains many more branches: the structure keeps repeating itself at different parts of the tree. For this reason we do not draw the whole tree but only part of it. This is provided in Figure 8.2. The tree starts with a random node corresponding to the first piece of paper being turned over, with its subsequent result, X_1 say. Clearly it does not matter, at any stage, which piece

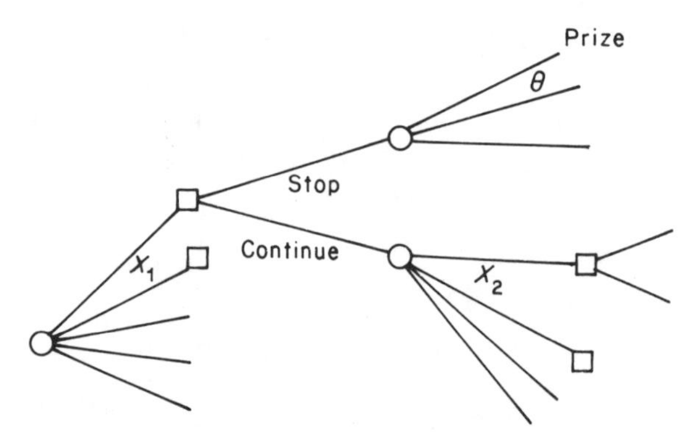

Figure 8.2. Decision tree for the marriage problem

is inverted. For every possible result there is a decision to be taken: either to stop and declare X_1 to be the largest number, or to continue. In the figure this is only shown for a single X_1; similar branches would follow every other value. In the former case the true value, θ, of the largest number is declared and the prize, if any, awarded. (θ is an uncertain event, hence the notation.) Otherwise a second piece is turned over and the result X_2 obtained. The tree continues in this way until the final piece is inverted when the decision to stop is unavoidable.

There are no difficulties over the utilities, being defined by the prizes. The probabilities too are easily found. Since the prizes depend only on the ranks of the number, that is, whether a number is highest, second highest, or so on, only the probabilities of the different ranks need be considered. Thus at the first inversion the probability of turning over the highest number is $1/N$, the same value holding for all the other numbers. The probabilities at later stages are only a little more complicated: thus if the third number is the largest so far seen then the probability that it is truly the largest of all N numbers is $3/N$. This is easily found using the laws of probability; the details are omitted. With these and similar probabilities inserted, the decision problem may be solved by starting at the last stage, when the final piece of paper is inverted, and proceeding back through N stages until the issue at the first inversion can be settled. The calculations are not difficult but are a little tedious and are omitted. We content ourselves with summarizing the main results.

Consider first the version where a prize is only given if player two is correct. Then it is clear that it is no use stopping unless the number just revealed is the largest seen so far; for if it is not, it cannot be the largest and no prize can possibly result from stopping, whereas there is still hope from continuing. Consequently the only cases where it could be worth stopping are where the current number is the largest out of those so far seen, the exception being at the last stage, where stopping is inevitable. Suppose, under these circumstances, R pieces have been seen. Then it is intuitively reasonable that if it is worth stopping for some value of R it is certainly worth stopping for any greater value of R. For if being best out of R is good enough, being best out of some larger number must be an even better reason for stopping. This intuitively sensible result is confirmed by calculation, and hence it remains only to find the least value of R, R_0 say, for which it is worth stopping. The optimum strategy is then to turn up R_0 pieces of paper without stopping, and then stop whenever that last, or any subsequent number, is the greatest so far seen. Calculation shows that for N large the value of R_0 is about one third of N (more precisely, 0.368 times N), the approximation being quite good for N as low as 10. The result is reasonable from a commonsense viewpoint since it seems natural to have a look at a few pieces of paper to see what sort of numbers are being used before committing oneself to a final conclusion. Many people are surprised that as high a proportion as one third should be inverted before even venturing to stop. The expected value of this version of the game is about one third of the prize. There is only a probability of 0.368 of being right even with the optimum strategy, so that the game is heavily weighted in favour of the player who selects the numbers.

In the version where a prize of k dollars is awarded if the selected number has rank $N-k$ (so that a prize of as much as $N-1$ dollars is possible) it still turns out that one should turn up about one third of the pieces before committing oneself. Thereafter, however, the strategy differs because there is some point in stopping even when the current number is not the best so far seen. It is possible by using an optimum strategy to get an expectation quite close to the maximum possible prize.

8.8 SOCIOLOGICAL APPLICATION

A sociological situation where such a model contains some of the elements of the real-life phenomena is provided by the important decision problem of when to make a proposal of marriage. Lady readers may care to interchange the sexes in what follows. A gentleman, in the course of his life, will meet several ladies and some of them he will pay court to and eventually contemplate marriage. With each such lady there comes a point where he must either make a proposal of marriage or pass to a fresh companion. After having seen several he may decide he should have proposed to one he knew earlier, but there can be no going back, for she is now engaged elsewhere. He must propose to the current 'girl friend', or drop her and pass to the next. The similarities with 'googol' are evident: the ladies replace the pieces of paper, the inversion is the courtship, and the decision to stop is a proposal of marriage. To complete the analogy it must be supposed that the total number, N, of ladies is known (though we saw that some aspects of the solution depended but little on N) and that proposals are invariably accepted. The question of utility remains to be considered.

The 'googol' problem is therefore sometimes known as the 'marriage' problem, or as the 'secretary' problem, for similar situations arise in choosing a secretary. Before readers dismiss the model as being too ludicrous for words, let me remind them how different from reality is the astronomer's model of the solar system that he uses to predict the motions of the planets. Similar gross approximations may be legitimate for some purposes in sociology as well as astronomy. The model here has some elements that are correct: the sequence of ladies, the difficulty of proposing to any but the current companion, and the learning by experience that accompanies one through life. It may well be that the grosser conclusions that follow from the model may have some relevance in real life.

We saw that with 'googol', for both the utility functions, it was best not to stop until quite a substantial proportion, say around one third, of the pieces of paper had been inspected. A similar conclusion will apply to the model for the marriage problem: namely that a gentleman should pay court to several ladies before venturing to propose to one of them. The exact point where he contemplates proposals will depend on the utilities, but since, as we saw with 'googol', the point seems fairly insensitive to variations in the desirability of being married, for example, to the second-best, the rule seems of rather general applicability. (This is an example of what was meant above by one of the grosser results of the analysis.) The model therefore argues in favour of rather later marriages than are customary nowadays. For example, suppose this process works from the age of 18 to 40; that is, over 22 years. Then the first third of the period should be spent on inspection, so that not until about 26 should a gentleman contemplate a proposal. This is at variance with current practice. There are at least two explanations: either people are not coherent or the model is grossly wrong. The two explanations are not exclusive and both are probably correct. The example is inserted here not as a serious contribution to sociology but as an indication of the way in which one might attempt to structure one of life's important decision problems. In the form of 'googol' it remains one of the few really complicated decision problems that has been solved in any sort of generality as distinct from specific numerical instances. In any case I fail to see why such approaches should not be able to make as valuable a contribution to these important sociological problems as much that nowadays passes for valid argument amongst the literate, but scarcely numerate, intelligentsia. More hopefully, a proper blend of both types of skill might be expected to produce even more satisfactory accounts of sensible behaviour.

8.9 TWO-ARMED BANDIT

The next example is again of a simple model with possible applications to the conduct of human affairs. There are many situations in which it is possible to repeat a decision problem, the repetitions being identical apart from the knowledge gained by appreciating the consequences of one's actions. For example, a motorist may have two routes he can take to work and has to choose between them. He has to make this choice every working day, the problem remaining the same apart from the experience gained from the journeys on previous occasions. At the beginning he has little knowledge of the respective merits of the two routes and it seems natural for him to try both and thereby gain experience of them: later, as he acquires experience, one route may seem clearly the better and then he will adhere to that choice. Our example is designed to show how this procedure may be correct from a coherent viewpoint such as we are adopting in this book.

Our model is usually called a 'two-armed bandit'. A one-armed bandit is essentially a machine with one arm which, on being pulled, can either result in a win or a loss. In Britain it is often called a fruit machine, because success depends on the haphazard arrangement of some fruits. It is supposed that on each pull of the handle there is the same chance of success, and that successive pulls do not influence one another (though they may influence your beliefs about them). A two-armed bandit is a similar machine but with two arms: the arms have nothing to do with each other, and the chances of winning are not necessarily the same for the two arms. Essentially it is two one-armed bandits.

The problem with a two-armed bandit is which arm to pull. If only one pull is allowed the problem is straightforward, but if a sequence of pulls is envisaged the situation is none too clear.

Consider, for example, the case where one is fairly sure about one of the arms, to the extent of being almost certain that the chance of success with it is 1/2, but much less knowledgeable about the other; though, if forced to make a judgement, it would be that it has not got as high a chance of success as the first arm. If a series of pulls is envisaged it may be worth trying a few goes at the second arm, even though it is probably worse than the other simply

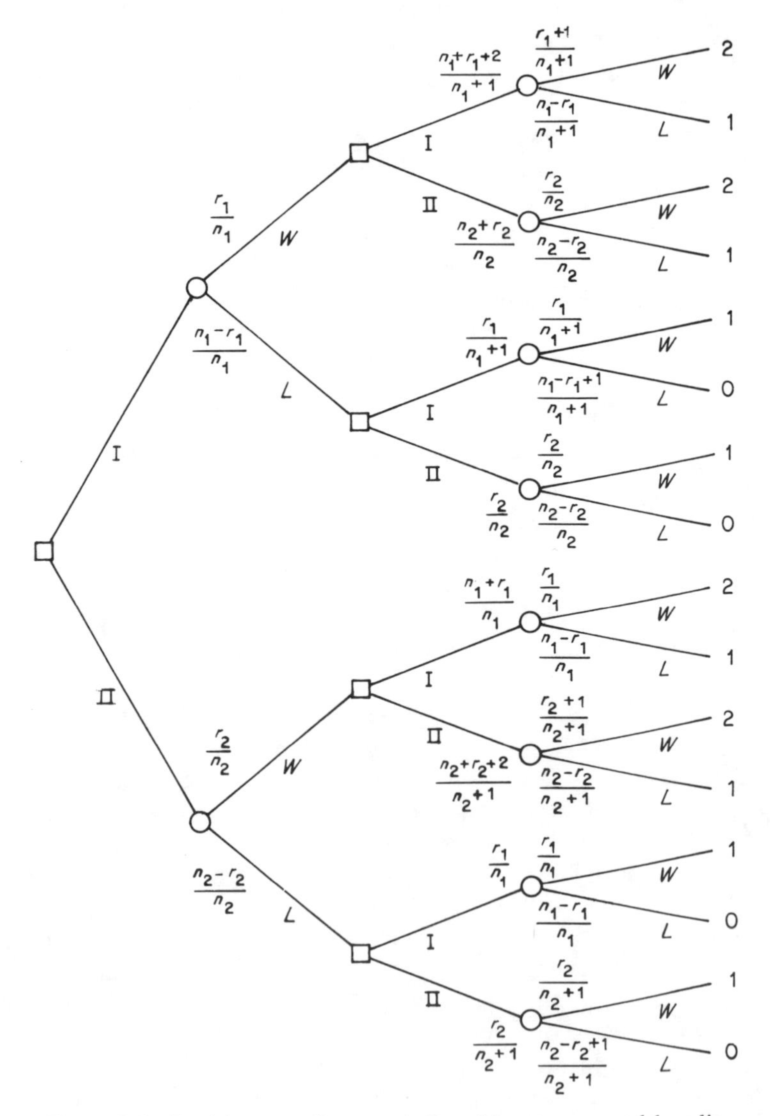

Figure 8.3. Decision tree for two pulls with a two-armed bandit

because one knows so little about it. In the same way the motorist might try a strange route, even if not very promising, simply because his knowledge of it is vague and it might, after all, be good. We want to show that such exploratory investigation of apparently poor acts is sensible within our framework. To do this for a reasonable number of pulls with the two-armed bandit leads to tedious calculations: we therefore confine ourselves to two pulls, with the benefits of simple arithmetic and easier comprehension, but the snag that the advantage of the exploratory strategy will not be very large.

The complete decision tree for two pulls of the two-armed bandit is shown in Figure 8.3. At the base there is a decision node with two branches corresponding to the selection of arm I or arm II for the first pull. Following, for example, that corresponding to I, we reach a random node with two branches labelled W, for a win as a result of pulling the first arm, and L, for a loss. At the end of each of these branches there is another decision node, again with two branches, representing the selection of the arm for the second pull. Either I or II leads to a random node with branches for the resulting W or L. The other branches lead to similar structures. The utilities are easily inserted: we suppose each win gives a prize of one unit, each loss preserves the *status quo*. (A loss of -1 would provide the same results but introduce the added complexity of negative values.) As a result each terminal branch has utility of 2 (both wins), 1 (a win and a loss), or 0 (both losses).

8.10 PROBABILITY CALCULATIONS

The probabilities need a little more care and to simplify the analysis we use the technical device mentioned in section 6.15. Here the probability of success is described by two numbers r and n: their ratio is the probability immediately associated with a win, and n measures how sure one feels about that assessment. These numbers change according to the two rules: on a win both r and n increase by 1; on a loss, r is unaltered but n increases. The probabilities can now be inserted on to the branches emanating from random nodes. It is supposed that initially arm I has values r_1 and n_1, and arm II, r_2 and n_2. If I is selected for the first pull the probability of a win is r_1/n_1, of a loss $(n_1 - r_1)/n_1$. Suppose a win results and I is again selected for the second pull. The probability of a win now increases to $(r_1 + 1)/(n_1 + 1)$. Had II been selected the probability of a win would be r_2/n_2, the original value since that is the first pull of that arm. Other probabilities can now be inserted in the tree. The results are shown in Figure 8.3 and some, at least, should be verified by the reader.

The tree is now labelled with the utilities and probabilities and the analysis, stage (4), can now be completed by alternate expectations and minimizations starting at the tips of the tree. At the top right-hand corner of Figure 8.3 corresponding to two pulls of I, with the first resulting in a win, we have a chance $(r_1 + 1)/(n_1 + 1)$ of 2 and a complementary chance $(n_1 - r_1)/(n_1 + 1)$ of 1: hence the expectation is

$$2(r_1 + 1)/(n_1 + 1) + (n_1 - r_1)/(n_1 + 1) = (n_1 + r_1 + 2)/(n_1 + 1)$$

and this value is attached to the random node. We may proceed similarly with the other random nodes to the right of the figure—that is, at the tips of the tree. For example, if arm II is first tried, results in a loss, and, as a consequence, arm I is tried for the second pull, the expectation is

$$r_1/n_1 + 0 \times (n_1 - r_1)/n_1 = r_1/n_1$$

It is now possible to go back to the decision nodes corresponding to the second pull. Again taking the uppermost one in the figure we see that I has an expectation of $(n_1 + r_1 + 2)/(n_1 + 1)$, where II has $(n_2 + r_2)/n_2$. The better act is to select the arm having the higher value. This decision and its expected value are not inserted into the figure because their determinations require numerical values for r_1, n_1, r_2, and n_2; but the principle should by now be clear.

8.11 NUMERICAL EXAMPLE

We next select values for these quantities, taking

$$r_1 = 1,\ n_1 = 2.2;\ r_2 = 10,\ n_2 = 20$$

Here $r_2/n_2 = 1/2$ with $n_2 = 20$, so one is fairly sure that arm II has a 50% chance of success: even a failure with it will only reduce one's assessment of this chance to $r_2/(n_2 + 1) = 10/21$, about 48%. On the other hand, arm I has a lower chance, namely $r_1/n_1 = 1/2.2$, about 45%, and one is much less informed about it: the low value of n_1 meaning that a single failure with it will reduce the assessment to 1/3.2, only 31%; on the other hand, a success will raise it to 2/3.2, about 62%.

Figure 8.4 repeats Figure 8.3 with these numerical values except that the terminal branches of the earlier diagram have merely been indicated and their details, already investigated, have been omitted. Thus the result of the first pull with I having been a win, the expected value with a second pull of the same arm is $(n_1 + r_1 + 2)/(n_1 + 1)$, as calculated above; this gives 5.2/3.2 in our case and this value has been placed alongside the appropriate random node in Figure 8.4. The expectations of all the possible second pulls can similarly be found and are inserted against the random nodes to the right of the diagram. (The values have been left as fractions, the more easily to appreciate their derivation.) Now pass back to the decision nodes referring to the second pull. The decisions associated with each of these can be selected by choosing the branch with the greater expected reward. Thus if I had been used at first and resulted in a win (the upper of the four nodes) it is better to continue with I with expectation 5.2/3.2, as against II with expectation 30/20. On the other hand, if the first pull of I had resulted in a loss (the next lower node) it is better to change to II, expectation 10/20, rather than persist with I, whose expectation, because of the loss and its subsequent serious effect on one's opinion about it, is only 1/3.2. The decisions, when arm II had been pulled first, follow similarly.

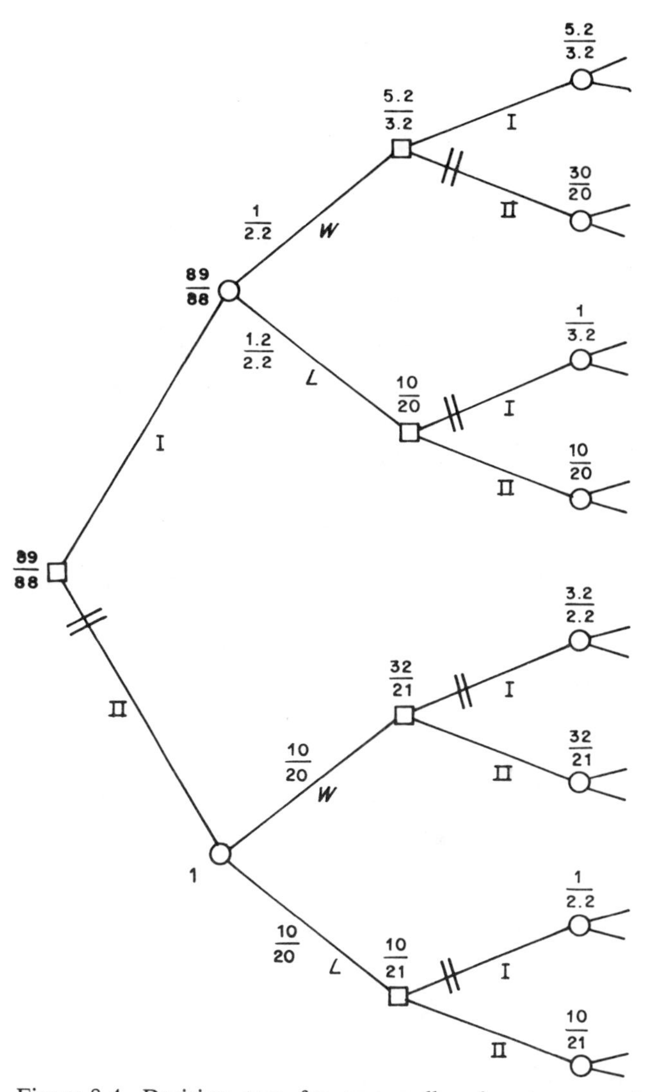

Figure 8.4. Decision tree for two pulls of a two-armed bandit (numerical case)

A further move down the tree takes us to the random nodes corresponding to the first pull. The probabilities are transferred from Figure 8.3 and the rewards associated with the outcomes have just been found, consequently the expectations at these nodes are easily calculated. If I had been used first we have

$$\frac{1}{2.2} \times \frac{5.2}{3.2} + \frac{1.2}{2.2} \times \frac{10}{20} = \frac{89}{88}$$

and for II

$$\frac{10}{20} \times \frac{32}{21} + \frac{10}{20} \times \frac{10}{21} = 1$$

Hence I has the greater expectation and it is better to use it for the first pull.

The two-move, two-armed bandit problem is now solved. In our numerical case where initially arm I has a 45% chance of success and arm II 50%, but our knowledge of I is weaker than II, it is better first to pull arm I and if it wins, pull it again; but if it loses to revert to arm II. In this way the expected gain is slightly higher (89/88) than if arm II is used on both pulls (1.0). Thus it is better to try an apparently inferior possibility because it just might not be as bad as you think. If strategy refers to the long-term decision problem with two pulls, and tactics apply to single pulls, then good tactics does not make for good strategy.

The numerical values in Figure 8.4 are worth pursuing further. We leave this to the reader and suggest he might like to use other numbers to see what happens. It is worth emphasizing how every part of the tree has to be investigated even if it is ultimately removed. For example, the whole of the lower half had to be calculated although it subsequently turns out to be irrelevant. It is an unfortunate fact that one cannot prune without having first inspected the branch to see if it has good possibilities.

The principle of the two-armed bandit for any number of pulls should now be clear, but the computation for any interesting number rapidly becomes enormous. At any stage one has to carry essentially four numbers along with one: r_1, n_1, r_2, and n_2. These change according to the rules already described and the calculator, human or electronic, has to keep account of them. The problem is mathematically one in four dimensions and no suitable algorithm seems to be available. It is worth noting how an apparently simple decision problem like this can rapidly get prohibitively large; which is one reason why this book does not contain any solutions to some of life's major decision problems.

8.12 THE MOUNTAIN-PASS EXAMPLE

We end this chapter by discussing a decision problem with a simple tree structure and hence a simple analytical solution; but one in which the probabilities and the utilities are harder to assess because of the non-quantitative nature of the problem in its original formulation. The example has been discussed before in sections 1.9, 3.7, and 6.9. It concerns a traveller having to cross a mountain chain in winter, trying to decide whether to use his car or travel by train. The unknowns of the situation concern the state of the pass over the mountains, the possibilities of an accident, and the chances of arriving in time for his meeting. The tree structure is shown in Figure 8.5.

There are only two decisions, to go by car or take the train, so the tree begins with a decision node having two branches. If the train is selected, then the

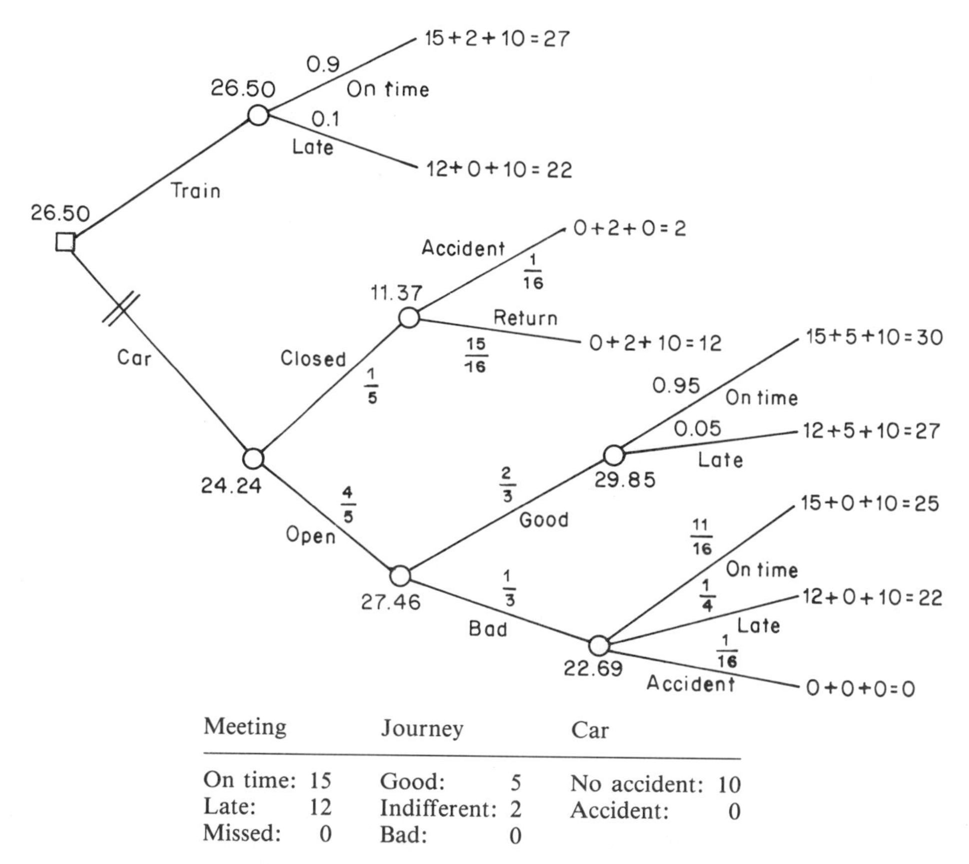

Meeting		Journey		Car	
On time:	15	Good:	5	No accident:	10
Late:	12	Indifferent:	2	Accident:	0
Missed:	0	Bad:	0		

Figure 8.5. Decision tree for the mountain-pass example

journey is either completed on time, or near enough so, and the appointment is made; or the train is so late that the decision-maker arrives late at the meeting. We ignore the possibility of his not getting there at all by this method on the grounds that this event has very small chance and can therefore be omitted.

Consider next the branch corresponding to the decision to go by car. The random node here has a complicated structure, due to the many possibilities, and it is convenient to break it down into a number of separate random nodes. The first dichotomy concerns whether or not the pass is open, the two branches being labelled 'open' and 'closed', respectively. Pursuing the latter one first, it divides into two according to whether the motorist is able to return to base unharmed—it being supposed he has then missed the train—or meets with an accident. If the pass is open, then it is first relevant whether the road conditions are good or bad—giving another random node—and if good, whether he gets there on time or not. If bad, we suppose there is not only the possibility of delay but also of an accident, so that this random node has three branches. The whole of the tree after the decision branch involving the car could have been condensed to a single random node with seven branches, but the tree given is simpler for the probability calculations. This completes the tree. We do not pretend to have introduced every possibility, but this problem arose in a real situation and there the other considerations seemed so unlikely to occur that they could, like the train not getting through at all, be omitted.

Consider next the probability structure of the tree. This has been discussed in section 3.7 and we describe it by means of a series of conditional probabilities. Pursuing first the branches corresponding to car travel, it is supposed that the probability of the pass being open is 4/5, 80%. One can imagine the factors that will have entered into this assessment: the snow that fell yesterday, the lorry you saw this morning thick with snow, the optimistic chatter of an extrovert neighbour, and so on. If open, the probability of the weather conditions being good is supposed 2/3, leaving only 1/3 to the possibility of the road being only open in one lane and driving both hazardous and slow. Finally, under the good conditions the probability of being late is as low as 1/20, whereas under the bad conditions it is as high as 1/4 and there is an appreciable chance, 1 in 16, of an accident resulting in missing the meeting altogether. Notice how these probabilities have been assigned in sequence so that any complete description has its probability evaluated by the product law: thus the probability of being late with bad weather conditions over an open pass is

$$p(\text{open}) \times p(\text{bad} \mid \text{open}) \times p(\text{late} \mid \text{open and bad})$$

that is, $4/5 \times 1/3 \times 1/4$, 1 in 15.

Returning to the possibility of a blocked pass, a probability, again 1 in 16, has been assigned to the possibility of an accident. Finally, from the decision to take the train there is a 1 in 10 probability of the train being late. A seasoned traveller on the line might be able to use statistical ideas to obtain this value, basing it on his many past experiences on the line and using Bayes' theorem. This completes the probability specification of the problem.

8.13 UTILITIES

The utility specification provides some new problems for us because we have no monetary values to guide us. I suppose that the traveller will have his expenses paid whatever route he uses, and that the car is adequately insured, so that all the pleasures (such as travel through snow-covered mountains in the sun) and pains (like an accident) are qualitative and not monetary. What has to be done is to associate with each terminal branch, or consequence, a utility describing, on a probability scale, the merits of the various outcomes. There seem to be three elements present in the consequences: first, the time of arrival at the meeting; second, the quality of the journey; third, the possibility of an accident. I have supposed the first at basically three levels: on time, late, and missing it completely, Similarly, the second has three levels, good, indifferent, and bad. The final feature either contains an accident or not. Then to a first approximation any consequence can be described in these three 'dimensions'.

Thus the train arriving late gives late arrival, a bad journey, but no accident. Next I have associated the utility zero with the worst of each of the three elements and other values with the best, these latter being chosen to reflect their importance. Thus, supposing the meeting to be a very important one, a value 15 is attached to an arrival on time, 5 to a good journey, and 10 to freedom from an accident. The intermediate possibility of being late has been given a value 12, and of a moderate journey 2. The values are listed in the table alongside Figure 8.5. Finally any consequence is supposed to have its utility assessed by adding the contribution of the three elements together. For example, arriving late with the train and classifying a journey in an unpunctual train as a bad journey gives $12 + 0 + 10 = 22$. Let us see what these assumptions imply.

The best consequence, arriving on time after a good journey free from accident, has utility $15 + 5 + 10 = 30$: the worst has utility zero. Thus the consequence just considered of arriving late on the train is equivalent to a chance of 22/30, about 73%, of the best consequence and a complementary chance of the worst. This is hard to appreciate, so let us look at it another way. The additivity of the three numbers means that the advantage of an improvement in one element is independent of the other elements. Thus arrival late instead of not at all, an improvement of 12, is the same whether the journey that led to it was good, bad, or indifferent. Bearing in mind the diminishing marginal utility of money you may feel that this assumption is false, and that an improvement is worth less in good conditions than in bad. We make the additivity assumption here because it is simple and seems to be reasonably near the truth. The three values associated with the best of each of the three elements, 15, 5, and 10, have been chosen to reflect the fact that getting to the meeting on time has three times the pleasure of a good journey, which in turn has only half the value of freedom from accident. Obviously the orders of magnitude are right, the precise values are hard to defend: we suggest below experimenting with variants of them. Suppose, for example, the consequence of arriving on time after a bad journey, free of accident is considered: the utility is $15 + 0 + 10 = 25$. Now suppose the consequence improves by the journey being good: call this C. Next, suppose it deteriorates by one missing the meeting: call this c. (The accident possibility is irrelevant in the immediate discussion.) Then the original consequence is intermediate between C and c: C having utility $15 + 5 + 10 = 30$, and c, $0 + 0 + 10 = 10$. Then missing the meeting is not compensated for by a pleasant journey. Indeed, a gamble on C and c would require odds of 3 to 1 on C before it was equivalent to the original consequence. ($3/4 \times 30 + 1/4 \times 10 = 25$.)

8.14 COHERENCE OF UTILITIES AND PROBABILITIES

The values inserted for the utilities and probabilities are in no sense correct and any other values wrong. They represent the decision-maker's individual preferences and may be modified by him. The only inviolate feature of them is their coherence. We have studied in some detail how this coherence is achieved with the probabilities within the problem by using the laws of probability. But one must also remember that there is coherence of this problem with other decision situations. For example, tomorrow I may be faced with the problem of where to take my children in order to get some good sledging. If the circumstances have not changed, then I should still adhere to the chance of 1/5 of the pass being blocked, a chance that is clearly relevant to the availability of good sledging conditions. If the conditions have changed then the value of 1/5 is still relevant because it will be needed in a proper use of Bayes' theorem, taking into account the altered circumstances.

Similarly, the utility values must cohere with related quantities in other decision problems. Thus I may, on some future occasion, consider going into the mountains in order to admire the scenery and walk on the ridges. I will again have to balance the pleasures to be derived from these activities against the hazards of driving with difficult road conditions and the possibility of an accident. It would be absurd for me in those circumstances to assign a utility of less than 5 to the pleasures to be gained from the excursion, since these are certainly not less (for me) than the delights of a good car journey. A decision problem in isolation can have any values for utilities and probabilities. It is the coherence with other problems that constrains their values.

In Figure 8.5 the utilities associated with the consequences have been expressed as the sum of three numbers in the way just described. One can now work backwards down the tree in the way previously discussed. Most of the nodes in the present tree are random nodes, so several expectations have to be found. These have been entered against their corresponding nodes in the tree. Finally, this procedure takes us back to the original decision node, where we see that the train journey has an expected value of 26.50 whereas the car yields only 24.24, and the decision is pretty clearly in favour of the train. The reason for this is essentially the high chance of not being able to make the meeting if the car is selected, the advantages of the car in improving the pleasantness of the journey not being enough to compensate. We leave the reader to verify that the chance of the pass being blocked would have to drop to 6% before the car becomes the preferred mode of travel. The reader is again invited to try other values for himself or even to include additional features in the tree.

Exercises

8.1. Rework Exercise 7.5(i) using a decision-tree.

8.2. Verify the statement made towards the end of this chapter that, in the mountain-pass example, Figure 8.5, the chance of the pass being blocked would have to drop to 6% before the car becomes the preferred mode of travel.

8.3. A company has to decide whether or not to drill for oil in a particular spot. It costs c units to make a seismic test, the result of which will be 'good', 'fair', or 'bad' prospect of oil. The actual drilling operation costs 75 units. There are three possible results of drilling: a high yield of oil which can be sold for 200 units, a moderate yield of 100 units, or no oil. The company's data for previous places of this type are as follows:

	High	Moderate	None	Totals
Good	20	10	10	40
Fair	9	9	12	30
Bad	3	12	15	30

In addition, drilling was not carried out at places with the following seismic records:

Good: 0, Fair: 10, Bad: 20.

Had drilling been carried out it is believed that the results would have been similar to those in the present table. In the past, in places of this type, a seismic test has always been carried out. What is the maximum value of c to make a seismic test worthwhile? If the actual seismic cost is 5 units less than this value, determine the optimum interpretation of the test in terms of whether or not to drill. What is the expected profit?

8.4. A part of an aircraft engine can be given a test before installation. The test has only a 75% chance of revealing a defect if it is present, and the same chance of passing a sound part. Whether or not the part has been tested it may undergo an expensive rework operation which is certain to produce a part free from defects. If a defective part is installed in the engine the loss is L (utiles). If the rework operation costs $L/5$ utiles and 1 in 8 of parts are initially defective, calculate how much you could pay for the test and determine all the optimum decisions.

8.5. A doctor has the task of deciding whether or not to carry out a dangerous operation on a person suspected of suffering from a disease. If he has the disease and does operate, the chance of recovery is only 50%: without an operation the similar chance is only 1 in 20. On the other hand, if he does not have the disease and the operation is performed there is 1 chance in 5 of his dying as a result of the operation, whereas there is no chance of death without the operation. Advise the doctor. (You may assume there are always only two possibilities, death or recovery.) 4/13

8.6. Rework the previous exercise where recovery can now mean either complete recovery or partial recovery. If the operation is performed and recovery results it is always complete. If the operation is not performed and recovery results it will be partial if he has the disease and complete otherwise. Advise the doctor. (You may assume that the operation cannot be performed later on if the patient lives.) $4/(14-\lambda)$

0 = death, 1 = full recovery, λ = partial recovery

Chapter 9

The assessment of probabilities and utilities

'If only he could have understood the doctor's jargon ... so as to be sure he was weighing the chances properly ... And yet he *must* decide!'

In Chancery, Ch. 12.

9.1 THE NEED FOR ASSESSMENTS OF PROBABILITY AND UTILITY

The argument used in this book has essentially been a logical one that if you wish to behave sensibly, for example, to use the sure-thing principle (section 3.1), then the logic of the decision situation forces upon you the consideration of probabilities and utilities and the principle of maximization of expected utility. Or, to put it the other way round, to use any other decision-making procedure would lay one open to the charge of doing absurd things, like losing money for sure in a Dutch book (section 3.15). A criticism of the argument says that logic is all very well but people are not logical and could not be, so that the procedures suggested are impractical. Now it is undoubtably true that people are not logical. Pyschologists have performed numerous experiments in simple decision-making that almost all show illogical behaviour by the subjects. So that part of the objection is admitted. The second part, that people could not be logical, is more open to discussion. The question is: accepting that people are not logical and do not think in terms of probabilities and utilities, could they not be trained to think that way? If they could not, then our thesis remains an abstract ideal; but if they could, then there is the potential for an improvement in decision-making over the illogicalities currently pursued.

The only way to answer that question is to train people to think probabilistically and see whether the training is successful. At the moment people cannot easily, if at all, assess the probability that a government will be re-elected, or that a currency will rise. But by instruction they might find it possible and worthwhile to do it. We discussed in section 4.8 reasons why it would seem sensible to attempt to get people to think in terms of probabilities and utilities. Essentially, it would be as absurd to dismiss this theory because the measurements are difficult as it would have been to have dismissed Newtonian mechanics at the end of the seventeenth century.

9.2 STATISTICAL EVENTS

Let us therefore address the problem of measuring probabilities and utilities. It is unfortunately true that at the moment we know very little about how to do this. Indeed, one of the reasons for writing this book is to explain the logic and the properties of the approach in the hope that people will be sufficiently impressed to try the method out for themselves and therefore assess utilities and probabilities. We begin with probabilities, later discussing utilities. There are some events, usually called statistical (section 2.3), whose probabilities, or chances, are readily available. The chance that an ordinary coin will fall heads when reasonably tossed, the chance that a normal British female aged 65 will die before reaching her eightieth birthday, are immediate or can be obtained from actuarial tables.

There are also chances that are not so readily available. An example is provided by the famous birthday problem. There are 23 unrelated people in a room: what is the probability that two of them celebrate their birthdays on the same day, the year of birth being irrelevant? With 365 days in the year there seems plenty of scope for the 23 people all to have different birthdays and most people respond to the question by giving a small value. In fact, the correct probability is slightly over 1/2. Notice that this is an example of coherence. All would agree that the chance of being born on a particular day is 1/365 (excluding leap years) and is unaffected by an unrelated, neighbour's birthdate. It is then a consequence of the rules of probability that the chance exceeds 1/2. What is at issue here is the probabilistic calculation that people typically cannot do; in default they assess unnecessarily many values and do not, because they cannot, check for coherence.

(As an aside, it might be noted that there are experts in probability but these people have been trained in the use of the rules of probability and not in the assessments of the values. They are like geometers who have never measured the angles of a triangle in their lives.)

So there are statistical events whose probabilities can be assessed or calculated from others. There remain probabilities that are not statistical and for which an extensive body of experience is not available to provide an agreed value. How are these to be assessed? As said above, nothing like a complete answer is available but three tools exist: scoring rules, coherence, and indirect methods. These are discussed in turn.

9.3 ASSESSMENT USING SCORING RULES

Scoring rules were introduced in section 2.10, motivated by the desire expressed in section 2.9 to measure how good a probability was, and subsequently used to justify the rules of probability. An example discussed in section

2.11 concerned the event of 'rain tomorrow' in a particular place. The reader was encouraged to try his hand at some questions and see how the introduction of the rule affected his assessments.

A scoring rule does not so much improve the assessment of a single probability but rather provides one with general experience in the assessment problem that should be of value in any special case. The example of training by experience that is usually quoted is that of meteorologists repeatedly assessing the probability of 'rain tomorrow'. The hope is that they will be better assessors as a result of experience with scoring rules. The more ambitious idea is to include scoring rules as part of everyone's education so that children would learn to appreciate and measure uncertainty, and not be forced to answer 'yes' or 'no' to a question about which they are genuinely uncertain. A European child, uncertain about the capital of the United States and asked 'Is New York the capital?' would not have to respond with a hesitant 'No' but could respond 'Probability 0.2'. Hopefully, this would train people to think probabilistically and to face uncertainty as a real feature of this world and not to ignore it or even deny its existence by inventing gods who know it all. In view of the potential importance of scoring rules, let us explore their use a little more closely.

We saw in section 2.12 how a scoring rule encouraged one to adopt a frequency or chance as a probability. In other words, it did a good measurement job with statistical events. This idea generalizes. Suppose that you have an event whose probability is truly P and that you are required to assess it by a value p of your choice. Your score will either be $(p-1)^2$ if the event is true, of probability P, or p^2 if false, probability $(1-P)$. The expected score (section 4.7) if you state p is therefore

$$\begin{aligned}&(p-1)^2P+p^2(1-P)\\&=p^2P-2pP+P+p^2-p^2P\\&=(p-P)^2+P(1-P)\end{aligned}$$

Treating the score as a negative utility, the prescribed rule is to maximize expected utility or minimize the expected score. The second term $P(1-P)$ does not depend on p and is unaffected by the selection of p. The first term is least, at zero, when $p=P$. Consequently your expected score is least when p is chosen as P. The quadratic scoring rule works because it encourages honesty: if P is the true value, you will announce $p=P$. We say it is *proper*. Consequently it works for all events, not just statistical ones.

9.4 ALTERNATIVES TO THE QUADRATIC RULE

The attentive reader will by now have asked himself the question: why use the quadratic scoring rule; what about other rules? Suppose the penalties had been different, would one still have assessed p as P? More importantly, would another rule have led to probabilities at all? Perhaps some other methodology would have emerged to replace maximization of expected utility. To answer these questions notice that the quadratic rule works because it encourages honesty: it selects p as P, it is proper. There are many other proper rules. For example, the logarithmic rule with penalty scores $-\log p$ if the event is true and $-\log(1-p)$ if false is also proper,* $\log p$ being the logarithm of p. At the moment we do not know whether one proper rule is better in some sense than another. This can only be investigated by trying rules out on subjects. Readers with access to logarithms might like to re-do the questions of Tables 2.1 and 2.3 with the logarithmic rule and see what difference it makes. The principal difference between the quadratic and logarithmic rules is that with the former the penalty never exceeds 1, whereas arbitrarily large values are possible with the latter. These can arise when p is near 0 or 1, and so the logarithmic rule perhaps discourages these extreme values more than does the quadratic.

9.5 IMPROPER RULES: ODDS

There are rules that are improper. These are of two types. As an example of the first and useful type consider a square root rule that scores $1/\sqrt{p}$ if the event is true and $\sqrt{p}$ if false, for stated probability p for the event. Here $\sqrt{p}$ means the square root of p and p is not restricted to the unit interval but can be any positive number. As before, the expected score is

$$P/\sqrt{p} + (1-P)\sqrt{p}$$

Write $\sqrt{p} = x^2$ so that x has to be chosen. The expectation is

$$P/x^2 + (1-P)x^2 = [\sqrt{P}/x - \sqrt{(1-P)}x]^2 + 2\sqrt{P(1-P)}$$

Again only the first term matters and is least, at zero, when

$$\sqrt{P}/x = \sqrt{(1-P)}x$$

or

$$x^2 = \sqrt{(P/(1-P))}$$

or

$$p = P/(1-P)$$

But $P/(1-P)$ is the odds on the event. So this rule would encourage people to state the odds rather than the probability. This would make perfectly good sense. Instead of using probability, everything in this book could have been written in terms of odds. Some results would have been harder to appreciate, some easier. For example, consider the addition law (section 3.2)

$$p(E_1 \text{ or } E_2) = p(E_1) + p(E_2)$$

* Readers familiar with the differential calculus will easily see that the expected score has zero derivative when p is P and that this gives a minimum.

In terms of odds, it reads

$$0(E_1 \text{ or } E_2) = \frac{0(E_1) + 0(E_2) + 20(E_1)0(E_2)}{1 - 0(E_1)0(E_2)}$$

which is decidedly more complicated. The multiplication law (section 3.5)

$$p(E_1 \text{ and } E_2) = p(E_1)p(E_2 \mid E_1)$$

becomes in terms of odds against $0^*(E) = p(\bar{E})/p(E)$

$$0^*(E_1 \text{ and } E_2) = 0^*(E_1)0^*(E_2 \mid E_1) + 0^*(E_1) + 0^*(E_2 \mid E_1)$$

only a little more complicated. We already saw in section 3.11 that Bayes' theorem is easier in odds form.

Bookmakers commonly use odds against and are familiar with the odds form of the multiplication law just quoted. If a horse in one race is at odds 2–1, and a horse in a second at 3–1, they will immediately give odds of 11–1 that both horses win. But suppose the horses are running in the same race at these odds, then the bookmaker will typically have difficulty in quoting odds for a bet that pays if either horse wins, because he needs the complicated odds form of the addition law. The coherent odds are 5–7 against. Of course, bookmakers are not, in our sense, coherent; they want to win. This is most easily seen by taking the odds for all the horses in a race, converting them to probabilities, and adding the results. The sum will be less than one. A Dutch book (section 3.15) cannot be made against them because they will not accept bets on $\bar{E}$, that is, bets on a horse losing.

Returning to alternatives to the quadratic rule. There are many rules that encourage honesty, not about a probability but about some function of probability, such as odds. These might be better if people were to find it easier to think in terms of the other function, but the basic laws are simplest when expressed in terms of probability and so probability is most likely to be the preferred description. Only experiments with people can settle the issue. There is a lot to be said for the logarithm of odds which has some attractive technical properties. In a sense bookmakers use the logarithm, for if you take typical odds they are finely stated for low values but coarsely expressed for high ones. Thus we get 11–7, 5–2, but rarely anything between 100–1 and 125–1. This difference would be greatly diminished on transferring to logarithms.

9.6 OTHER IMPROPER RULES

There is a second type of improper scoring rule that is unsatisfactory. As an example, consider a score which is the distance of the stated p from 1 if the event is true, and from 0 if false. In other words, it is the same as the quadratic rule without the squaring: $1 - p$, or p. As before, the expected score is

$$\begin{aligned} &(1 - p)P + p(1 - P) \\ &= p(1 - 2P) + P \end{aligned}$$

Again, only the first term matters. If $1 - 2P$ is positive the minimum is at $p = 0$, if negative, at $p = 1$. Thus if $P < 1/2$, say $p = 0$, if $P > 1/2$, say $p = 1$; either p being acceptable if $P = 1/2$. This is an absurd rule, for it can only produce two values, 0 or 1, according to whether the event is thought to be less or more likely to happen than not. Absurd it may be, but it is this type of rule that is implicitly used at the moment when a child is forced to answer 'yes' ($p = 1$) or 'no' ($p = 0$) to a question concerning a matter about which he is uncertain, like the capital of the United States (section 9.3).

In summary, there are three types of scoring rule.

(1) *Proper for probability*, like the quadratic or logarithmic rules;
(2) *Proper for a function of probability*, like the square root rule which is proper for odds, but improper for probability;
(3) *Totally improper*, giving only two values, like the distance rule.

The reader may be puzzled by the continued use of expected scores when discussing improper rules, because the notion of expectation was derived from the proper quadratic rule (section 4.3) and might reasonably not be valid in general. In fact it is, and we have used the expectation concept for simplicity in exposition. The reader who wishes can carry through the calculations for the other rules using only the sure-thing principle.

9.7 COMMENTS ON SCORING RULES

We have seen that a scoring rule is a potentially useful device in training people to make probability assessments. It is not known whether some rules are better than others, nor whether probability, rather than an alternative form like odds, is the preferred method of thought.

It is often said that scoring rules are defective because they can only be used if the event's truth can subsequently be established. Thus we only have to wait until tomorrow to find out about the rain but there is no way of telling whether Shakespeare truly did write the plays that are attributed to him, so that scoring rules are useless concerning the uncertainty of the authorship of *Hamlet*.

The criticism is not really important, because if an uncertain event has not consequent events that will, or will not, happen within a small time-span the original event does not matter. Does it materially matter who wrote the plays? Of course, if evidence should come to light to show that Shakespeare did not write them and the Earl of Oxford did, then Stratford will suffer an economic decline and Lavenham prosper. Perhaps Stratford could take out insurance against the event, but the probability is of a verifiable event, namely, 'determination, within the policy period, that Shakespeare did not write the plays'.

Similar difficulties arise in science. Scientists refer to theories but they cannot easily determine whether they are true or false, so probabilities of theories cannot be assessed by scoring rules. But good theories have verifiable consequences that can be tested by experiments and are hence amenable to scoring rules.

The difficulty of non-verifiability also occurs in the problem of whether to drill for oil at a specified place. A geologist will study the evidence and report on the uncertain event that significant amounts of oil are available there. By our thesis he should report the probability. The event will only be verified if the company he is advising decides to drill: for if not, the oil situation there will remain unknown. Yet the decision to drill will partly depend on the geologist's stated probability. The difficulty can be lessened by his stating the probability, were the company to drill, they would find oil, with no score if no drilling takes place. But there are still problems due to the interaction between the probability and the verification.

Another difficulty with scoring rules is that a person may be using an implicit rule that is improper for probability, so that although he appears to quote his probability he is not doing so. Here is an example using a modification of the quadratic rule. Suppose he assesses a loss when the event is false to be twice as serious as when it is true. This is in imitation of the geologist, who will be in grave difficulties if the oil is not present and drilling takes place. Suppose, then, the score is $2p^2$ if the event is false and $(1-p)^2$ if true. The expected score is

$$(1-p)^2P + 2p^2(1-P)$$

for true probability P. This is

$$P - 2pP + p^2P + 2p^2 - 2p^2P$$
$$= (2-P)(p - P/(2-P))^2 + P - P^2/(2-P)$$

and is least when $p = P/(2-P)$, which is always less than P. Hence the geologist will underestimate the probability.

Implicit rules can often cause trouble. An engineer who has developed a new machine will have a scoring rule that implicitly reflects his enthusiasm for his 'baby' and will overstate the probability that the machine will function. One can guess at his scoring rule and correct accordingly. For example, if it was surmised that the geologist was using the modified quadratic rule just mentioned, stated value p could be converted into $2p/(1+p)$ and used as the proper probability. ($p = P/(2-P)$ becomes $P = 2p/(1+p)$.)

9.8 COHERENCE

Use of a scoring rule demands nothing more than the selection of a single number to describe an event. Yet probability, or odds, or whatever form the rule uses, obeys the laws which get forgotten during this assessment, simply because they are not needed. This is a major defect of scoring rules: they do not exploit the laws of probability, they do not use coherence. A surveyor wishing to measure the distance between two points typically does not just measure the single distance, he measures other distances and uses triangulation (the laws of geometry) to obtain a better determination of the distance required. The same idea can be used with advantage in assessing probabilities.

In thinking about one uncertain event, introduce others, assess their probabilities as well, and use the laws of probability (instead of the surveyor's geometry) to obtain a better assessment of the original probability. With judicious choice of the other events, all the laws of probability can be invoked and full coherence achieved. This is perhaps the best practical way of evaluating probabilities; by thinking about a topic, rather than a single event, and thinking coherently about it.

This idea has already been mentioned briefly in connection with the mountain-pass example in section 3.7. Let us consider it again in the fuller form of the example in section 8.12, the probabilities being stated on the decision tree in Figure 8.5. The following notation is used:

C: the car was used,
O: the pass was open,
G: road conditions were good,
A: an accident,
T: arrival on time.

The tree gives the following values:

$$p(T \mid \bar{C}) = 0.9,\ p(O \mid C) = 4/5,\ p(G \mid C, O) = 2/3,\ p(A \mid C, \bar{O}) = 1/16,$$
$$p(T \mid C, O, G) = 0.95,\ p(T \mid C, O, \bar{G}) = 11/16 \text{ and } p(A \mid C, O, \bar{G}) = 1/16:$$

whilst some others have been supposed zero. Here, as usual, a bar over a letter denotes the negation of the event to which the letter refers: thus $\bar{C}$ means going by train (not car). Notice that all these probabilities have been assessed for events in the natural time order, first C, then O, G, A, and finally T; the probability for any event being assessed is given the earlier events in the sequence, thus, A given C, O, $\bar{G}$. This is usually the most satisfactory method, partly because it is easiest to think that way with the present influenced by the past, but also because the individual probabilities can take any values (between 0 and 1) irrespective of the other probabilities. There are seven numbers given above: these could coherently have any values. However, having selected these values (and some zeroes) *all* the other probabilities in the problem are determined. There is no choice left and the rules of probability enforce all the other values on you. Let us see how this works.

9.9 PROBABILITY CALCULATIONS

Some implications are obvious. Thus what is the probability of not having an accident given the car was selected, the pass open, but conditions bad? In the notation, $p(\bar{A} \mid C, O, \bar{G})$. By the addition law this is $1 - p(A \mid C, O, \bar{G}) = 15/16$. Similarly, the probability of arriving late under the same circumstances is $1 - 11/16 - 1/16 = 1/4$. There are more subtle implications that use the extension of the conversation, and hence the multiplication law. For example, given that the car is used, what is the probability of an accident? By extending the conversation to include the state of the pass, we have

$$p(A \mid C) = p(A \mid O, C)p(O \mid C) + p(A \mid \bar{O}, C)p(\bar{O} \mid C)$$

$$= \frac{1}{3} \times \frac{1}{16} \times \frac{4}{5} + \frac{1}{16} \times \frac{1}{5} = 0.029$$

The second term is straightforward, the first also uses the road conditions on the open pass; only if these are bad (1/3) is an accident probable (1/16). This value of 0.029, or odds of about 33–1 against, is a direct consequence of the other assessments. If the decision-maker is unhappy with it, then some of the earlier probabilities must be revised.

Yet further implications are provided by Bayes' theorem with its reversal of uncertain and conditioning events (section 3.10) and hence here a reversal of time-order. For example, suppose the decision-maker learns that a neighbour has had an accident on using the pass: what is his probability that the pass was open? We require $p(O \mid A, C)$, which may be written $p(O \mid A)$, the use of the car being understood. By Bayes' theorem this is

$$p(O \mid A) = p(A \mid O)p(O)/p(A)$$

$$= \frac{1}{3} \times \frac{1}{16} \times \frac{4}{5} \Big/ \left(\frac{1}{3} \times \frac{1}{16} \times \frac{4}{5} + \frac{1}{16} \times \frac{1}{5}\right)$$

$$= \frac{4}{7}$$

the denominator having just been calculated. Thus the news of the accident has reduced the probability of the pass being open from 4/5 to 4/7. As a further example, the reader might like to calculate the probability that road conditions are good on learning that the neighbour, using his car, reached the meeting on time. The value is 0.734.

In the decision tree (Figure 8.5) the alternatives car or train were choices leading to a decision node. In another view they could be random, so leading to further calculations. For example, the decision-maker receives a telephone call from his neighbour telling him that the neighbour has reached the meeting on time: what is the probability that the neighbour used the car? We need $p(C \mid T)$. With Bayes' theorem in odds form,

$$O(C \mid T) = \frac{p(T \mid C)}{p(T \mid \bar{C})} O(C)$$

The two probabilities are easily found by considering the various ways that the meeting can be reached on time. Directly, $p(T \mid \bar{C}) = 0.9$ and $p(T \mid C) = 0.95 \times 2/3 \times 4/5 + 11/16 \times 1/3 \times 4/5 = 69/100$. The ratio of the latter to the former is 23/30, so that the odds on the neighbour having used the car have decreased by this factor. For example, if the decision-maker originally had odds of 3–1 on his neighbour using the car, the call has reduced them to 23–10, a little over 2–1 on. If his assessed reaction to such a call is different from this, then some revision is necessary in the original probabilities. The lesson always is that coherence forces arrangements of opinions according to the inviolate laws of probability.

9.10 ELECTION FORECASTING

This book is really all about coherence, and coherence must always be the major tool in understanding uncertainty. So let us consider a second example of using coherence to assess probabilities. The primary uncertain event is whether, in an election to take place in six months' time, the government will be re-elected; an event R. This may be assessed directly as $p(R)$, omitting the general conditions H from the notation. The assessment of $p(R)$ does not use coherence, except trivially in obeying the convexity law and making it lie between 0 and 1. To employ coherence, other events need to be introduced even if they are not of primary interest. We need to think about the electoral process rather than just its final outcome. So which events should be considered? Clearly, they should be related to the primary event: it is no use including irrelevant ones.* People are strongly influenced by their own concerns and will support the government if they are doing well, which in turn depends on the strength of the economy. So let us introduce the event E that the economy is strong in the run-up to the election. (One could consider some measure of the strength but our concern here is with principles, not calculation details.) Another factor influencing people is the threat of war and the involvement of their military in it. So consider the event W of a war in a neighbouring country.

We have two events, E and W, that influence re-election R. It is usually easier to consider probabilities like $p(R \mid E, W)$, rather than $p(R)$ directly, because the conditions (of a strong economy with a war) give a more specific scenario in which to think. Essentially, the conversation is being extended from R to include E and W. In fact, the choice of additional events should be influenced by the type of scenarios it is easiest to contemplate. The inclusion of two events leads to four scenarios, and possible assessments are

$$p(R \mid E, \bar{W}) = 0.9$$

$$p(R \mid E, W) = 0.8$$

$$p(R \mid \bar{E}, \bar{W}) = 0.5 \qquad (9.1)$$

and

$$p(R \mid \bar{E}, W) = 0.4$$

These values reflect the powerful effect of a strong economy, increasing the probability of re-election by 0.4 both with and without a war. The occurrence

* It is sometimes hard to be sure what is relevant. There is a case of a government of a country arguably being defeated because the country's football team had been beaten in the final of the World's Cup the week before the election.

of a war has a similar, but opposite, effect of reducing the probability by 0.1 both in strong and weak economies.

To complete the study of the election it is necessary to include a discussion of the additional events, E and W. In capitalist societies the occurrence of a war boosts the economy, especially if the war is not on one's own territory, by creating a demand for weapons. The assessments

$$p(E \mid W) = 0.8$$

and

$$p(E \mid \bar{W}) = 0.5$$

reflect this, the war increasing the probability of a strong economy by 0.3. Finally, suppose war has a probability of 0.3 ($p(W) = 0.3$). All the probabilities have now been assessed; seven instead of the original one, but some of which are easier to think about, and all of which are helpful in thinking about the election.

The original $p(R)$ can now be evaluated by extending the conversation to include the economy and the war status. There are four terms in the sum (section 3.8) corresponding to the four scenarios. The first is

$$p(R \mid E, \bar{W})p(E, \bar{W})$$

which, on applying the product law to the second probability, gives

$$p(R \mid E, \bar{W})p(E \mid \bar{W})p(\bar{W})$$

Inserting the numerical values, this is $0.9 \times 0.5 \times 0.7 = 0.315$. Similarly, the other three terms, in the order of (9.1), are 0.192, 0.175, and 0.024, giving a total of 0.706 for the probability of re-election.

Other probabilities can, and should, be evaluated as checks on coherence and aids in understanding the situation. For example, the probability of a strong economy $p(E)$ is seen, by extending the conversation to include war, to be

$$p(E \mid W)p(W) + p(E \mid \bar{W})p(\bar{W})$$

$$= 0.8 \times 0.3 + 0.5 \times 0.7 = 0.59$$

It may seem obvious that the effect of a war would be to lessen the probability of re-election because (equation (9.1)) in both states of the economy it has that effect. So consider $p(R \mid W)$. Again by extension of the conversation, this time to include the economy,

$$p(R \mid W) = p(R \mid E, W)p(E \mid W) + p(R \mid \bar{E}, W)p(\bar{E} \mid W)$$

(Notice that in this use of the law all probabilities are conditional on W: in effect, W is part of H.) Inserting the numbers we easily have $p(R \mid W) = 0.72$, and similar calculations with no war yield $p(R \mid \bar{W}) = 0.70$. Consequently re-election is slightly more probable with a war than without one, despite the fact that it is less probable in each economic state. This is another example of Simpson's paradox (section 3.12). The explanation is that although war has a damaging effect in either state of the economy, it can have such an effect on strengthening the economy, and that in turn is so powerful in securing re-election, that the overall effect is beneficial to the government. The example has sinister overtones. If the government has control over whether a war starts it may well decide to start one because that will increase its probability of re-election. It could do this at the same time as it emphasized to the electorate the damaging effect of a war in any state of the economy. Clever governments do not always need to lie, though they often do.

Having investigated the dependence of R on W, consider that on E. Extending the conversation to include W we have

$$p(R \mid E) = p(R \mid E, W)p(W \mid E) + p(R \mid E, \bar{W})p(\bar{W} \mid E)$$

and this requires $p(W \mid E)$. This last is not a natural probability to contemplate for it suggests that the economy influences war, rather than war encouraging the economy. However, it is available by Bayes' theorem,

$$p(W \mid E) \propto p(E \mid W)p(W) = 0.8 \times 0.3 = 0.24$$

and

$$p(\bar{W} \mid E) \propto p(E \mid \bar{W})p(\bar{W}) = 0.5 \times 0.7 = 0.35$$

so that on normalizing to add to 1 they are 0.407 and 0.593. Hence $p(R \mid E) = 0.86$. Similar calculations for $\bar{E}$ give $p(W \mid \bar{E}) = 0.146$ and $p(R \mid \bar{E}) = 0.49$. The powerful effect of a strong economy is confirmed, increasing the probability of re-election by 0.37.

It is possible to evaluate other probabilities like $p(W \mid R, \bar{E})$, the probability of a war given the government was re-elected in a weak economy, but they have little interest or relevant meaning except perhaps to someone who has not read overseas news, confining his attention to domestic issues. Bayes' theorem would be required but has little relevance because the time-reversal it uses is only of value when a past event is uncertain (as with the guilt of a defendant at law).

9.11 THE ASSESSMENT OF SMALL PROBABILITIES

It is often necessary to consider events that have a very small probability of occurrence because the consequences, were they to occur, would be serious. An accident at a nuclear power station provides an example. Here new problems arise connected with the tiny values involved. It is hard for us to appreciate what a probability of 1 in 1000 means, let alone one in a million. Which is one reason why log-odds may be preferable (section 9.5) here giving values 3 and 6, respectively (to base 10).

A useful device is to compare the event with another that is more familiar and has about the same probability. For example, 1 in 1000 is a little more probable than four 6's in four tosses of a fair die (1 in 1296). In a recent investigation concerning nuclear waste the increased probability of leukaemia was said to be about the same as the increased probability of death as a result of using one's private car rather than public transport. The implication here, of course, is that you accept the latter so why not the former? The implication is quite unjustified: the statement is merely a comparison of probabilities, not utilities. Nevertheless, the comparison does give one some sort of feel for the hazard.

The most frequently used method of assessing a small probability uses the multiplication rule (section 3.5). If the basic event only occurs if several other events occur we can think of it as 'E_1 and E_2 ... and E_n' and evaluate its probability as

$$p(E_1)p(E_2 \mid E_1) \ldots p(E_n \mid E_1 \text{ and } E_2 \ldots \text{ and } E_{n-1})$$

For example, a nuclear accident only occurs if the core overheats (E_1), the alarm mechanism fails (E_2), and so on. Each of the probabilities in the product must be larger than the product and more reasonably than the basic event avoid the difficulties associated with very small values. For example, three probabilities each of the understandable value of 1 in 10 give the low 1 in 1000.

The method has two snags. First, it is easy to think that the event can only happen in one way. Often this is not so, and several possibilities need consideration. These have to be added, if exclusive, to reach the final evaluation. Consequently it is easy to underestimate the small probability when using the multiplication law. The second snag is that is tempting to think of the constituent events E_1, E_2, ... E_n as independent (section 3.14) and use the simpler expression $p(E_1)p(E_2) \ldots p(E_n)$. This can be dangerous, since complicated interactions can cause events to occur that independence does not allow for. A famous court case, much discussed in the literature, involved a statistician making judgements of independence that were only weakly supported and having his testimony rejected because of this.

9.12 INDIRECT METHODS

People are faced with uncertain situations every day of their lives and make decisions, despite the uncertainty. Sometimes it is possible to infer their probabilities indirectly from the actions they take. For example, insurers, especially when dealing with non-statistical events, assess a premium without consideration of probabilities. From this action it should, in principle, be possible to say something about the uncertainty of the event being insured. There are difficulties: first, that people are typically not coherent; second, that the decision depends on the utilities as well as the probabilities, so that unless the former are known, the latter cannot be calculated. Consider a simple example.

You are offered a ticket which will award you a prize of 100 dollars if the government referred to in section 9.10 is re-elected; that is, contingent upon R. If it is defeated you get nothing. How much will you pay for the ticket? Many people will answer the question whilst refusing to contemplate a probability. If the response is, say, 40 dollars, it is tempting to infer $p(R) = 0.4$, since with assets C you can either gain 60 or lose 40, and this is equivalent to the *status quo* C if the expected return matches C, which happens with a probability of 0.4;

$$0.4 \times (C + 60) + 0.6 \times (C - 40) = C$$

However, the calculation is suspect because it supposes the utility for money is linear or even that money is the consequence you consider. Maybe you like the gamble and attach more value to $C + 60$ when winning 100 dollars than to a straight gift of 60 dollars.

The situation need not be as artificial as this example. I once met a chemical engineer who realized that the process for which he was responsible might fail but was reluctant to assess the probability of failure. However, he knew the monetary consequences of failure (and they were essentially all that mattered), so I asked him: suppose I were able to offer you a device that would be sure to make it work, how much would you pay me for it, 1000 dollars, 10 000? He laughed at the latter figure but contemplated the former more seriously. After some discussion he settled for 750, from which, as in the last paragraph, his probability could be found. But again, his utility for money was supposed linear.

Indirect methods have the advantage that they can adapt to the method of thinking that the decision-maker finds natural. This is important, because people have developed skills by thinking in a particular way and it is good to exploit these skills instead of forcing them into a probabilistic framework that they find unnatural and where they cannot use their experience. Nevertheless, reactions to decision problems involve utilities and probabilities, and without the first, the second cannot be found. Let us therefore turn to the assessment of utilities.

9.13 THE UTILITY OF MONEY

We begin with the simplest case where the consequences are entirely monetary so that no other considerations, such as the pleasures of gambling, arise. There are two ways of assessing the utility: fixed-state and fixed-probability methods.

Fixed-state method

The subject is asked to consider that he has initial capital (or assets) C and is invited to compare this with a gamble in which he might either lose L or gain G, reaching assets $C - L$ or $C + G$ (L and G are both positive); in particular, what probability p of gain (and $1 - p$ of loss) would make him indifferent between C and the gamble.

Fixed-probability method

This is the same except that p is fixed, G not, and the subject is asked to select the value of G that secures indifference.

We shall only discuss the fixed-state method because it appears to have some advantages over the other, especially in generalizations to other utility problems, but many of our remarks equally apply to the fixed-probability method. First, a few comments on the method. An alternative description has the subject put up a stake L to win G and have the stake returned (compare the Dutch book in section 3.15). The probability used is not an assessed probability of some event but one referred to a standard such as balls from an urn (section 2.5). Indeed, the question might be rephrased as how many white balls in the urn of 100 balls would give indifference, it being understood that white gives a gain and black a loss. So the argument is not circular: utility assessment does not require one of probability, but probability may, as we have seen, need one of utility.

Equating the expected utilities, the expressed indifference gives the equation

$$U(C) = pU(C + G) + (1 - p)U(C - L)$$

Since utility is unaffected by changes of origin or scale (section 4.10) the utility of the worst state (or consequence) $C - L$ may be taken as zero and that of the best $C + G$ as 1. Then $U(C) = p$ and $U(C)$ is determined.

9.14 COHERENT UTILITY

This determination nowhere invokes coherence, which must be our main tool, just as it was with probability. To invoke coherence, write $C - L$, C and $C + G$ as C_1, C_2, and C_3, respectively, and introduce a further state C_4 better than C_3. Then, in addition to the question about indifference between C_2 and a gamble on C_1 and C_3, yielding p_2 (previously p), the subject can be asked to compare C_3 with a gamble on C_2 and C_4, yielding p_3, say. (In the latter the loss is $C_3 - C_2$, the gain is $C_4 - C_3$.) With $U(C_1) = 0$ and $U(C_4) = 1$, the two indifferences yield the equations

$$U(C_2) = p_2 U(C_3)$$

and

$$U(C_3) = p_3 + (1 - p_3)U(C_2)$$

with solutions

$$U(C_2) = p_2 p_3/(1 - p_2 + p_2 p_3)$$

and

$$U(C_3) = p_3/(1 - p_2 + p_2 p_3)$$

Thus four utilities have been found and coherence is invoked by asking for a further comparison of C_2 with a gamble on C_1 and C_4 which, in replacing C_3 by C_4, is different from, yet of the same character as, the previous comparisons. (The gain is $C_4 - C_2$.) If p is the indifference probability stated

$$U(C_2) = p$$

(as $U(C_1) = 0$, $U(C_4) = 1$) and hence the coherent answer to the last indifference assessment is

$$p = p_2p_3/(1 - p_2 + p_2p_3)$$

Any other value than this one would lead to a Dutch book being made against the subject in the manner of section 3.15.

The general procedure is to select n asset positions $C_1, C_2, \ldots C_n$, each larger than the one before, and embracing the monetary range (C_1, C_n) required; to put $U(C_1) = 0$, $U(C_n) = 1$ and to ask for comparisons of C_i with a gamble on (C_{i-1}, C_{i+1}) for $i = 2, 3, \ldots n - 1$. The stated probabilities determine all the utilities $U(C_i)$ and coherence can be investigated using comparisons of C_j with a gamble on (C_i, C_k), $i < j < k$, having loss $C_j - C_i$ and gain $C_k - C_j$. There are very many of these and a thorough verification of coherence is possible. Section 5.8 provides an example.

It is possible to use more complicated gambles involving more than two states. This is particularly useful if such a gamble arises naturally and the subject is used to thinking about it. For example, in an each-way bet on a horse-race there are two possible gains: one if the horse wins, another if it gains a place. This requires two probabilities but the complication may be a simplification if the subject is familiar with such bets. As with probability, it is always useful to present the person with situations that are natural for him, rather than contrived.

9.15 UTILITY OF A QUANTITY

The methods just developed for monetary consequences apply equally well to other consequences that are measured in terms of a single quantity: examples abound; units of energy available, amount of water resources, examination grades, travel times between two places. In fact, even less than a quantitative evaluation is needed, only an ordering. Here is an example of the latter.

Quality of eyesight is measured on a scale from zero to 10, zero corresponding to complete blindness and 10 to perfect sight. The intermediate numbers merely provide an ordering: thus 5 is better than any number less than 5 but is in no real sense halfway between blindness and perfection in the way that two lots of 5 dollars give 10. Yet the coherent methods just described work with 11 states. C_i corresponding to state i on the sight scale ($i = 0, 1, \ldots 10$). Consider a patient with bad eyesight in state 1 contemplating an eye operation that will most likely take him to state 6 but has a small probability of going wrong and making him totally blind, state zero. This involves a comparison between C_1 and a gamble on (C_0, C_6). Such a comparison may be more natural, at least to an ophthalmic surgeon, than one between C_5 and a gamble on (C_4, C_6) since the latter involves only small changes and would not be a normal form of surgery.

This particular example is of some interest because of the form of the utility function. With $U(C_0) = 0$ and $U(C_{10}) = 1$ as before, it happens that $U(C_1)$ is already quite large, say around 1/2, or even 2/3. This expresses the experience that even a small amount of eyesight is very valuable in that it enables one to get around with some aids. $U(C_5)$ is almost 1 because glasses can almost bring the subject's appreciation up to C_{10}. In particular, it follows that the operation in the previous paragraph would only be considered if the probability of total blindness, C_0, was very small.

One feature that is presumed in the methods of earlier sections is that utility increases with the value of the quantity. (Decrease, as with time of travel, would be equally acceptable, merely interchanging gain with loss.) If utility increases over one part of the range and decreases over another, care has to be exercised. An example is provided by the retailer faced with uncertain demand for his product (section 1.8). As before, denote by θ_j the state that the demand is for j items. If the store contains J items, then $j = J$ is optimum, for $j < J$ means unsold items and $j > J$ unsatisfied customers. The utility increases up to J and then decreases. Now some of the gambles suggested are inappropriate: for example, C_J against one on (C_{J-1}, C_{J+1}), since both changes from C_J are losses. However, other gambles can be substituted: for example, C_{J-1} against (C_J, C_{J+k}) at least for sufficiently large k. Here the gain, from $J-1$ to J, is an extra item sold; the loss involves a balance between unsatisfied customers and unsold items. There are sufficient possibilities to provide a check on coherence.

9.16 UTILITIES INVOLVING SEVERAL QUANTITIES

The same methods extend in principle, but with substantial difficulties in execution, from one to several quantities. We argue in the context of a decision problem of where to site a new airport for a city. Amongst the many quantities involved consider just two: the amount of pollution caused by the planes and the time to travel to the airport. (It is supposed that the questions of how these should be measured have been settled.) Denote these by s (stench) and t. Utility clearly decreases as either s or t increase since we want minimal pollution and a short journey. The available sites are either a long way from the city centre and cause little pollution or are near to the centre with a lot of people being polluted, so that large s goes with small t and vice versa. Figure 9.1 illustrates the situation and may be compared with Figure 5.4 and the discussion in section 5.13.

Now we require to measure the utility $U(s, t)$, so reducing the two original numbers, s and t, to one, U. If this is done there will be several pairs of values of s and t giving the same U and one will be indifferent between the pairs. Generally, there will be indifference curves in the plane of s and t having the property that you will be indifferent between any pairs lying on the same curve as the dots in the figure. A possible first task in constructing the utility function is to assess these indifference curves. This involves no utility measurement and no probability considerations but does mean a judgement of equality and of how much an increase in s balances a decrease in t. If an increase d in s is always compensated for by a fixed decrease in t the indifference curves will be

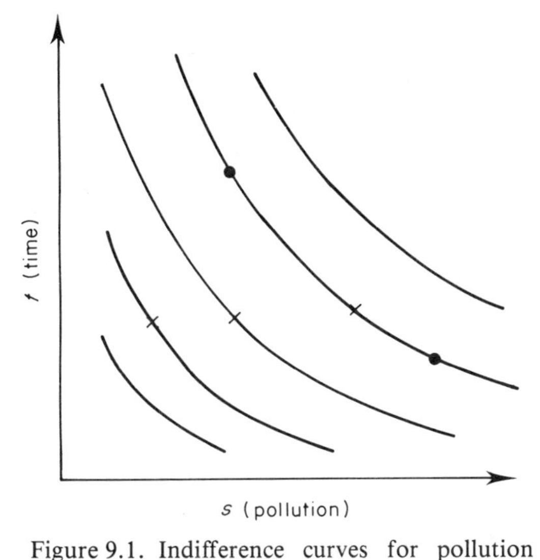

Figure 9.1. Indifference curves for pollution and time

straight lines running from top-left to bottom-right. Generally, they will flow in that general direction as shown in Figure 9.1.

Once the curves have been determined the problem reverts to the type already considered with only the utility value of each curve to be found. For example, three situations, shown by crosses in the figure, may be selected and the intermediate one contrasted with a gamble on the two outer ones. We have shown a case where the three have the same time value but different levels of pollution. One could interchange the roles of s and t or even consider cases where both variables change. There are even more possibilities than in the univariate case to consider for coherence. Besides coherence, attention must always be given to making the gambles meaningful and realistic to the person whose utility assessment is required. Thus the gamble corresponding to the crosses in the figure could be expressed in terms of a fixed site with a known time of travel and possible pollution levels due to changes in airplane design or uncertainties about the population densities.

These ideas extend from two to more quantities, but the number of gambles and the complexity of the indifference curves, or surfaces, rapidly increase. Simplifications are possible, but consideration of them would lead to technicalities such as independence, which are beyond the scope of this book.

9.17 GENERAL COMMENTS ON ASSESSMENT

The study of assessment, both of probabilities and utilities, is in its infancy. So little has been attempted and many of the attempts have been vitiated by the subjects' lack of familiarity with probability, thinking of it merely as a number between 0 and 1 and ignoring the addition and multiplication laws.

Probability, it has been said, is merely commonsense reduced to calculation. It is the basic tool for appreciating uncertainty, and uncertainty cannot be adequately handled without a knowledge of probability. It is even fundamental to the study of utility, for, as we have already emphasized, utility is not just a measure of worth, it is a measure of worth on a probability scale. It is this scale that enables the worth of consequences to be combined with the uncertainties of the same consequences in the form of expected utility, whose maximization gives the optimum decision.

Coherence is quite basic to probability and therefore to utility. Only assessment methods that employ coherence are likely to be successful, yet many papers in the literature on assessment fail to use the concept. We need more studies, firmly based on probability and coherence, before the practical success of the method of maximization of expected utility can be judged.

Chapter 10

An Appreciation

'. . . that means doing so long a sum every time that I can't think how you ever get to acting.'

Flowering Wilderness, Ch. 3.

10.1 THE LIMITATION TO A SINGLE DECISION-MAKER

The approach adopted throughout most of this book has so far been logical and rational; some would say unemotional and inhuman. We have asked whether there are any criteria that a person might reasonably want to apply to his actions, and have replied that he ought to be coherent both in his preferences amongst consequences and in his opinions about uncertain events. Granted these premises, which were defended in some detail, it has been proved by the laws of logic that the decision-maker possesses probabilities for the uncertain events, utilities for the consequences, and that the only sensible way for him to proceed is by maximizing expected utility. Some of the results of applying this criterion have been explored and found to yield sensible answers: for example, in portfolio selection (section 5.14).

In the present chapter the purpose is to survey the edifice that logic has erected and to consider both the advantages and disadvantages of maximization of expected utility. Our concern is not so much with how people actually make decisions, what was called the descriptive approach (section 1.4), but with how the normative, or prescriptive, view might contribute, or fail to contribute, to our lives. What are its strengths and its weaknesses? One weakness has already occupied us in Chapter 9, namely the difficulty in assessing probabilities and utilities, in describing things in numerate terms. The essential role of coherence was there emphasized and the need for training in numerate appreciation explained. The second important weakness is perhaps best described as a limitation. The theory only applies to a single decision-maker: it does not deal with the case of two or more decision-makers that often arises.

We have taken the situation of a single decision-maker, listing his courses of action, contemplating the relevant uncertain events, evaluating his probabilities and utilities, and calculating with those. What if two or more decision-makers are involved, each with his own array of probabilities and utilities, yet needing to agree on a single action? The situation is of common occurrence as when members of a committee have to make a decision; or when a community, be it a town or a country, has to make a choice between possibilities: and, most difficult of all, when countries have to take action in conflict with another country.

The first point to be made is that there does not exist a normative theory for several decision-makers. That is, it has not so far proved possible to write down a list of desirable qualities, like the coherent comparison of consequences, and deduce from these a procedure, like maximizing expected utility, that is logically optimum. When people are asked to cite outstanding, difficult, yet important problems, they often give replies like 'find a cure for cancer'. My own response would be 'find a theory of multiple decision-making'. Conflict is perhaps the most serious problem of our age because it threatens to destroy the world. Is war the only resolution?

There is, of course, a considerable body of literature on the multiple problem with more than one decision-maker, much of it useful. My point is that there is no solution available that even its inventors can logically defend. To use our overworked word: no coherent approach exists. Yet the need for one is more urgent now that at any point in the history of the world.

10.2 COMMITTEE DECISIONS

In default of an adequate solution let us see what light the one we have can shed on the multiple problem. We begin by considering the action of a committee that is required to reach a decision on a matter. A jury is a special case. Every member of that committee has his own views and might formalize them in the way we have suggested; but what if the conclusions differ amongst the members? First, notice that it is possible to think of the committee as a decision-maker, and for it to play the role of someone desirous of viewing the world coherently, of not violating the sure-thing principle, and hence of producing probabilities and utilities of its own. There is nothing in this book that requires the decision-maker to be a person: the theory concerns the choice of an action and the chooser could be a committee. There is just as much reason for a committee to be rational as for an individual. In other words, our problem can be thought of as passing from a set of values, one for each committee member, to a single valuation. How is this to be done? At present nobody knows.

We first argue that our approach has some value when applied to the committee problem, as distinct from the individual, even though it falls short of providing a complete resolution of the difficulty. The first task the individual has to do is to list the possible courses of action; the decisions d_1, d_2, ... d_m. Now it is surely sensible for the committee to do the same. Interaction between the members is capable of generating new decisions to add to the list.

10.3 COMPARISONS OF ACTIONS

One advantage in doing this that is often not understood is that it forces people into a comparison between acts rather than over-serious contemplation of single acts (see section 1.5). A course of action should be selected not because it is good but because it is better than other actions that are available. The concept of an absolute goodness is a chimera: only relative goodness is the reality that matters. The point arises with environmental issues. The environmentalists can see only the loss of animals and plants; the engineers appear only appreciative of the economic return. As a result, an antagonism develops that is totally unnecessary. If there are two decisions, to build the water reservoir or not, then both have advantages and disadvantages, and, like it or not, we have to effect a balance between that rare bird and the ability to water our crops in time of drought. It is perhaps our antagonistic legal system, with prosecution and defence who can see only bad and good of the defendant, respectively, that encourages this one-sided way of thinking.

The necessity for comparison also arises when there is no comparison. For example, a decision to build a new road may be vigorously resisted because of the destruction it will cause to the neighbourhood. However, it may happen that there is essentially no alternative to the new road, and if it is not built the whole economic basis of the area will collapse. The opponents of the new road must come up with an alternative proposal before serious discussion, involving comparison, is possible. It is absurd for them merely to criticize d_1 without providing d_2. Even the best of actions usually has unfortunate effects somewhere. A modification of the well-known proverb is apposite: 'it is a good wind turns none to bad.'

The enumeration of the courses of action available to the committee persuades its members to think of further possibilities. It is a defect of formal scientific thinking of the type used in this book that it does not include within the formalism any scope for originality or bright ideas. It is more the thinking of a computer than a human being. But here is an instance where the formalism calls for sparks of creativity. It does not suggest how they can be supplied, but it does show where they could be used. The most beneficial and rapid progress in this field, as in many others, will come through a blend of literary and scientific analyses. This listing of the decisions is just such an opportunity for the two approaches to come usefully together.

The theory then suggests that the committee should similarly enumerate the uncertain events. In less formal language, it should envisage what might happen. Even more than with the decisions, here is a case for creativity, for recognizing what might happen in the future, and what may be relevant for the actions the committee contemplates taking.

When people are asked to consider the future they often reply that this is impossible. Who could have foreseen the scale of the war in Vietnam? We cannot tell what technological advances will come about before the turn of the century. This is not asked of our scientific decision-maker. All that is required is that at the moment when he is considering how to act, he assesses, as best he can, the future events that will affect the outcomes of his actions. In no sense (as we shall discuss below) are these assessments to be regarded as correct, for the course of time will change their values. He has to look ahead as best he can, with the limited information at his disposal. Of course, if he could foresee the future, he could do better. The fact that we often are very uncertain about the future is one reason for difficulties in decision-making and for making information valuable.

10.4 PROBABILITIES AND UTILITIES

Suppose a committee has prepared the lists of actions and uncertain events; the theory then requires the assignment of probabilities and utilities. In Chapters 3 and 6 we have discussed the laws of probability and, in particular, have shown how Bayes' theorem enables probabilities to be changed by the proper incorporation of the information provided by data. These laws can be used with advantage in reaching agreement in committee over the chances of the uncertain events. For suppose two members contemplating an event θ assign different values $p_1(\theta)$ and $p_2(\theta)$. It has repeatedly been emphasized that all chances are conditional on the information available when the probability assessment is made, and therefore it may be that these two numbers are different because the members have (or are remembering) different information. Thus we might write, rather than $p_1(\theta)$ and $p_2(\theta)$, $p(\theta \mid A_1)$ and $p(\theta \mid A_2)$, where A_1 and A_2 are the two pieces of information the members possess. It would then be natural for them to communicate their knowledge one to another, when they will both possess A_1 and A_2. It is more reasonable to think that $p(\theta \mid A_1 \text{ and } A_2)$ might be the same for both members.

The pooling of information amongst members is clearly an important aspect of committee work. I see it as essentially a device for bringing people's probability assessments into reasonable agreement. However, it remains true to say that even with complete unity of experience people may not assign the same numerical values to the chances. The theory is silent about how the committee should resolve such a disagreement. Of course, no resolution may be necessary, for it may happen that one action has higher expected utility than the rest under either assessment. Or it may happen that members are prepared to modify their assessments enough to make this occur.

Similar disagreements may arise over the utilities. Here the situation is perhaps more serious than with the events because we have no rules of utility comparable with those for probability, and because there is definite reason to think that, even with the same information, people disagree over the desirability of consequences. The situation looks unpromising. Nevertheless the basic idea of coherence can make some contribution. The fundamental lesson of coherence is that ideas—in this case, consequences—should not be contemplated in isolation, but should be related to other ideas to see how they fit together or cohere. It follows that if a member presents opinions about consequences that are at variance with those of the rest of the committee, the chairman should see how these opinions fit with others the dissenting member has. For example, a member may be encouraging the board to enter a new market overseas. Exploration of how this fits with other actions may reveal that the basic reason for his attitude is nothing more than a desire on his part to visit that part of the world. Such an attitude is easily changed once it is exposed, or if not it will hardly command respect in the context of the directorate's decisions. So although the theory is deficient here, the basic concept of coherence has its uses.

10.5 GOVERNMENT DECISION-MAKING: DEMOCRACY

Consider next the situation where a group of people have collectively to take a decision but the numbers involved are too large for the group to form a committee: the citizens of a town or the inhabitants of a country. The usual method is to form a government and let that take the decisions. The government then acts roughly like a committee and the considerations just adduced apply: it should list the possibilities, the uncertainties, and, by pooling of information and coherence, hope to reach a conclusion. But there is the additional consideration of how the government is to be selected. The generally accepted method is by a democratic procedure, but what is democracy? The major difference between the Soviet Union and the United States is not between communism and democracy but over what democracy means. The countries of the Eastern bloc consider themselves democratic and see, in the difficulties of access of left-wing views to the media, lack of democracy in the West. So what is democracy? It seems amazing to me that the question is so little discussed: almost everyone talks as though it was a clear concept, yet it is not. Since democracy is a procedure whereby the preferences of individuals are translated into decision-making by governments, the question is relevant in making decisions and we might stay to consider it.

The question is a big one and we will consider just one aspect, the possibility of a coherent procedure whereby one passes from individual preferences to a collective preference. Here is a simple form of the problem. Each of a number of people considers the same set of decisions $d_1, d_2, \ldots d_m$ and each ranks them from the most to the least preferred. In our coherent view each person would have a set of expected utilities $\bar{u}_1, \bar{u}_2, \ldots \bar{u}_m$ to provide the ranking but the present point to be made concerns only the ranking and not the numbers. The problem is to pass from the complete set of rankings of the individuals to a single ranking appropriate to the group of people. Clearly, there are some logical requirements that should be imposed, having the same role as, say, the sure-thing principle for a single decision-maker. One apparently obvious requirement is that if one individual changes his ranking by preferring d_1 to d_2 when previously he judged d_2 better than d_1, everything else remaining unchanged, then if the original group ranking preferred d_1 to d_2, it should continue to do so. Roughly, if one person increases his view (of d_1), then the group should not reduce its view.

Now a surprising thing happens. With several 'obvious' requirements of this type, Arrow, in his famous impossibility theorem, showed that no collective ranking exists: there is no way of passing from individual to group preferences without violating some of the obvious requirements. An exaggerated statement of the result is that democracy cannot exist. The result is disturbing but can be criticized using the arguments above concerning comparison. Arrow shows that no really good method exists, but we want merely the best amongst available alternatives. This last problem is still unsolved. The question of what is democracy remains unanswered in this framework.

10.6 FREEDOM OF INFORMATION

Whilst we remain ignorant of how to choose a government our theory tells us something about how the government should work, namely by maximizing its expected utility. We do not know how to pass from the individuals' utilities to the corporate utility but we do know that the government, acting as a decision-maker, should act in accord with a utility function. One of the important advantages of our numerate approach is the ease of communication that it provides. Numbers are immediately intelligible: 2 is 2 is 2 and all know what is meant. Consequently governments could, and should, easily communicate to the electorate what utility function they are using. Indeed, they should do this at the election and the people could select their governments on the basis of the declared utility. How do you rate expenditure on defence against that on education; how much social welfare is advocated? This is done at the moment but less precisely than numeracy would allow. Furthermore, the present vagueness allows a government, when elected, to deviate from its promises; whereas a numerate expression would hinder this and the electorate could demonstrate the change in utility.

Probabilities also are of great value in communication. Governments often predict what will happen if their policies are pursued: inflation will come down to 4%. Readers who have understood this book will recognize this form of statement as a nonsense and a delusion. The future is uncertain, and the only proper statement about it is through probability: the probability that the rate will be 4% is 0.6. Again, because the statement is numerate, it can be communicated and checked. Coherence can be used to see whether the government is behaving sensibly.

Our discussion is necessarily brief but the point is that, although no logical process is known whereby individual views can be combined into a corporate view, the single decision-maker approach has something positive to say in its advocacy of coherence and its easing of communication. One thing is clear: the introduction of these ideas will be vigorously resisted by the establishment and existing governments because they will increase the control by the people of their leaders. This will happen because of the checks—you aren't using your declared utility—and because of the ease of communication. Governments love secrecy as a way of keeping power and will see any move to reduce it as a threat. So maximization of expected utility is a possible aid to open government. It will also have the effect of encouraging honesty. Capitalists, forced to expose their utilities, will find it less easy to pretend they are encouraging democracy in the Third World when in reality what they want is a market for their goods.

Information may also have a role to play in reducing government decision-making in favour of individual action. Here is an example that arose recently. A drug for the treatment of arthritis was found to have a serious side-effect, damaging the liver and ultimately killing the regular user. Out of the many thousands of users about 100 were thought to have died as a result of treatment by the drug. The government therefore banned it. An elderly lady wrote to a newspaper describing how, with the drug, she had been able to live a near-normal life, whereas without it she had been a cripple needing constant attention. She did not mind if it ultimately killed her, for 5 years free of pain was better than 10 years of continual agony. So surely the optimum thing is for the government to allow the drug to be available but to provide the information that this drug has probability 0.01 of leading to fatal liver damage. The individual can then use her utility function and decide whether to use the drug.

That was an example of where there was no need for a government to find a corporate expression but can leave the action to the individual. There are other instances of groups forcing things unncessarily. Religious bodies are prone to insist that all use their probabilities and utilities and refuse to allow dissent. Just because John is sure God exists, giving the uncertainty probability one and so violating Cromwell's rule (section 6.7), why should Jean have to agree? Because John opposes euthanasia, why should Jean not be free to decide for herself? Our theory appears to act in favour of individual freedom. Governmental action is required whenever there is potential or actual conflict between peoples, so let us discuss conflict.

10.7 CONFLICT

The most difficult type of decision problem involving two or more decision-makers arises when they are opposed. It can arise with individuals, or with committees (as with the boards of directors of rival firms), or, most seriously with rival governments. Formally we have the situation in which one decision-maker has one range of options d_1, d_2, . . . , the other has a different set d_1', d_2', . . . and the utilities of the consequences of pairs of actions (d_i, d_j') of one choosing d_i and the other d_j' are perceived quite differently. The simplest example is a game between two players, with d_i and d_j' their respective strategies, and where what one player loses, the other wins. There is a logical solution to this game, called minimax, that will be discussed in section 10.9, but no general solution to the conflict problem is known. There could be conflict situations with three or more decision-makers but experience shows that alliances soon form amongst some of the participants and the conflict becomes that between two. They will not therefore be considered further.

One possible way to proceed is as follows. For simplicity, let the two decision-makers be John and Jean. Then John is uncertain what choice d_j' Jean will take. He therefore assigns probabilities $p(d_j')$ that she will select d_j'. John can now select d_i as the decision of maximum expected utility to him. Jean can do the same, assigning $p(d_i)$ and using her utility. This is often sensible but has the difficulty that when John contemplates the probabilities of Jean's actions he has to look at it from Jean's view and therefore has to think about the probabilities she will assign to his actions, which themselves depend on her actions, and we are in an infinite regress. In other words, it is logically difficult to assign probabilities $p(d_j')$ or $p(d_i)$. The difficulty is not just one of assessment.

10.8 PRISONERS' DILEMMA

A famous conflict situation that illuminates the problem is the prisoners' dilemma; not to be confused with the puzzle of the three prisoners (section 3.9). John and Jean are both prisoners, unable to communicate. Each has available two strategies, to co-operate or not. The utilities are shown in Table 10.1, the first entry referring to John, whose choices are at the left; the second to Jean, with her strategies at the top. If they both co-operate they will both do better (1,1) than by both not co-operating (0,0). But if John co-operates whilst Jean does not, he will suffer and she will greatly benefit (−1,2), with the roles reversed in the opposite strategy (2, −1). Look at it from Jean's viewpoint. She is uncertain about what John will do and contemplates the probability that he will co-operate. But then she realizes that if he co-operates she will do best by non-co-operation, gaining 2 instead of 1: and if he does not she will still do best by non-co-operation, gaining zero instead of −1. So the probabilities are irrelevant, her utility is surely maximized by non-co-operation. Similarly John will not co-operate, so they will both have utility zero. But this is clearly not optimum because they each gain utility 1 by both co-operating. The method of maximizing expected utility would appear to have failed. But has it?

The difficulty with their both co-operating, which is not the maximum expected utility decision, is that it would pay either of them to deviate from it. For example, if Jean co-operates on the understanding that John will and then John lets her down at the last minute, she will drop from 1 to −1. Also it will pay John to deceive her, because by switching he will rise from 1 to 2.

Table 10.1 Prisoners' dilemma

Jean John	Co-operation	Non-co-operation
Co-operation	(1,1)	(−1,2)
Non-co-operation	(2, −1)	(0,0)

The solution of them both co-operating is highly unstable; whereas both not co-operating is stable, neither profiting from a departure from it.

In discussing the serious limitation of our arguments in only applying to the case of a single decision-maker we have seen that there is no adequate theory for the multiple case. Nevertheless our argument has a useful contribution to make, though it falls short of a completely satisfactory account because it fails to pass from individual to group preferences. Bearing in mind that it is comparative, not absolute, merit that matters let us look at other methods of decision-making that have been proposed and see how they fare in comparison with ours. The best-known rival candidate is the minimax method, which is often advocated in conflict situations.

10.9 MINIMAX

This method recognizes and uses utilities (or losses, see section 7.5) but not probabilities, either denying the existence of the latter save in special cases, or claiming that the method is to be used when they are unknown. Consequently in applying the minimax method we can still write out a decision table, with rows for the decisions and columns for the uncertain events, and insert the appropriate utilities, but are denied the probabilities. Table 10.2 gives a simple example involving two decisions and two events. Here u is a number greater than one. (u less than 1 is of no interest since in that case d_2 is certainly the better act.) The table is of a familiar form: d_2 corresponds to the safe investment certain to yield 1, d_1 is the gamble capable of giving a high yield u or of dropping to zero. The minimax method proceeds as follows.

For each row, that is, for each decision, select the *minimum* utility. In Table 10.2 this is zero for d_1 and 1 for d_2. Now choose that decision having the *maximum* of these values. In Table 10.2 this is d_2 with value 1. Hence the minimax method says choose d_2 and don't gamble. Generally with utilities u_{ij}, that d_i is chosen which maximizes the minimum (over j) of u_{ij}.

The method advocated in this book would assign probabilities, $p_i = p(\theta_i)$, and choose d_2 if $1 > up_1$; that is, if $u < 1/p_1$. Otherwise d_1 would be selected. Consequently the minimax method says never take the gamble, in contrast to our method which says take it if u is large enough. This failure of the minimax method to respond to the value of u is a major weakness. Would you refuse a gamble that might lose you 1 dollar (if θ_2 occurs) but would otherwise result in a win of a million dollars? Most of us would accept unless p_1 was very small.

Another criticism of the minimax method is that it does not take account

Table 10.2

	θ_1	θ_2
d_1	u	0
d_2	1	1

of information. Suppose we have the same situation as Table 10.2 and u has a modest value so that the minimax conclusion to select d_2 is not unreasonable. Now suppose the decision-maker is provided with information that makes θ_1 almost certain. He will still refuse the gamble despite the fact that he is nearly certain to profit from it. A good illustration of this phenomenon is provided by considering a piece of equipment that the decision-maker is about to use. Let θ_1 be that the equipment is sound, θ_2 that it is faulty. Let d_1 be the decision to use it, d_2 the decision to buy a new piece. Then however much testing of it is performed which shows it to be sound, the minimax recommendation is to buy the new piece and scrap the old.

10.10 INCOHERENCE OF MINIMAX

Although these criticisms of minimax have considerable force, if the argument of this book is worth anything we ought to be able to demonstrate that an adherent of minimax is incoherent. We proceed to do this. Coherence is a method of fitting things together properly, so let us see how the decision problem of Table 10.2 fits with that of Table 10.3. Here the events are unaltered and d_2 is still available, but there is a new decision, d_3, which is effectively the opposite gamble to d_1, winning when that one loses and vice versa. In Table 10.3 minimax would still select d_2.

Now put the two tables together so that d_1, d_2, and d_3 are all available. We have seen that by the minimax method d_2 is preferred to both d_1 and d_3. Consider a new act, d, formed as follows. A coin, believed to be fair, is tossed; if it falls heads, d_1 is selected, if tails, d_3 is chosen. What are the utilities of d for θ_1 and θ_2? Since the minimax operator believes in utility theory they are clearly both $u/2$. Consequently if u exceeds 2, d with utility of $u/2$ is preferred to d_2 with utility of 1. Consequently we have the situation where neither d_1 nor d_3 are liked, but a choice between them based on the toss of a coin is the best decision. This is incoherence.

Table 10.3

	θ_1	θ_2
d_3	0	u
d_2	1	1

10.11 MINIMAX WITH LOSSES

The minimax method is not usually applied to utilities in the manner just described but to losses (see section 7.5, where a loss, l_{ij}, was defined as the difference between the utility of the best act for θ_j and u_{ij}). Since losses are equal to a constant minus utilities all the maxima become minima and vice versa. Consequently the method chooses that d_i which *minimizes* the *maximum* (over j) of l_{ij}. It is this combination of the words in italics which, on abbreviation, gives the name *minimax*. Table 10.4 gives a simple example with the utilities on the left and the losses on the right. Thus for θ_1, d_1 is the best decision with a utility of 8 so that d_2, with a utility of 2 in that case, involves a loss of $8 - 2 = 6$. Applying minimax to the loss table gives, for d_1, a maximum of 4, and for d_2, one of 6. Hence d_1, having the smaller value, is selected. The force of some of the criticisms just applied are diminished in this alternative version, but a new difficulty arises.

To demonstrate incoherence, enlarge the situation by introducing a third decision d_3. This time we will add a new row to the original table, corresponding to d_3, in order to calculate the losses. The new forms are given in Table 10.5.

Notice that the losses for d_1 and d_2 when θ_2 obtains do not have the same values as before since, for θ_2, d_3 is now optimum, not d_2. The maxima for the three rows are respectively 7, 6, 7, so the best act is d_2 with a minimum of 6. Consequently the effect of introducing a new decision, d_3, is to change the preference of d_1 over d_2 into exactly the opposite preference. This is incoherent. Imagine, having decided to stay at home one evening rather than go to the theatre, you reverse the decision because you discover there is a concert available.

Consequently the minimax method, even its usual form applied to losses, is unacceptable. Even aside from these difficulties the minimax method seems unsound in principle because it ignores any assessment—not necessarily in numerical terms—of the uncertainties associated with the events. I am surprised that the method continues to be studied outside game theory, where is has a sensible meaning.

Table 10.4

	θ_1	θ_2
d_1	8	0
d_2	2	4

Utilities

	θ_1	θ_2
d_1	0	4
d_2	6	0

Losses

Table 10.5

	θ_1	θ_2
d_1	8	0
d_2	2	4
d_3	1	7

Utilities

	θ_1	θ_2
d_1	0	7
d_2	6	3
d_3	7	0

Losses

10.12 MINIMAX AND CONFLICT

There is one circumstance where minimax has something to recommend it and that is a conflict situation in which what one person wins, the other loses. It is usually referred to as a zero-sum, two-person game. To apply the minimax principle there it is necessary to introduce a mixed decision, that is, a decision in which one mixes, say d_1 and d_2, by applying d_1 with probability (or chance) p and d_2 with probability $1 - p$. We met an example in section 10.10 when mixing d_1 and d_3 to produce d, and also in section 5.2. Mixed decisions play no role in maximizing expected utility since they cannot have higher expected utility than that of the best decision amongst those mixed. In a conflict situation they could have merit in disguising from Jean what John is going to do. This is especially true in repetitions of the conflict or game.

A simple example is provided by the game of matching coins. John and Jean simultaneously expose a coin with either head or tail uppermost. John wins, getting both coins, if they match: Jean if they do not. Table 10.6 describes the two basic strategies of showing heads, and showing tails, together with the mixed strategy of choosing each exposure with probability 1/2. The utility of this is zero since the chances of $+1$ and -1 are equal, and it is clearly minimax. If the game is to be played repeatedly then the minimax procedure may be good and if Jean does something different John may be able to exploit it. For example, in play with young children they may alternate their choices more often than chance would dictate. If so one can easily win (or lose if you want to please the child) by guessing that they will do whatever they did not do last time. For a single play, maximizing expected utility is sensible and is effectively minimax for Jean if she thinks John is equally likely to use heads or tails.

Minimax is often advocated for other conflict situations and is regrettably often used by governments in 'defence' situations. There are several objections. First, the conflicts being discussed are not repeatedly played. Second, the situation is typically not zero-sum because the basic issue giving rise to the conflict is the different utilities of the decision-makers. Thus the United States and the Soviet Union have different utility functions. But the most important objection is that use of the minimax strategy fuels the arms race. In minimizing *her* maximum loss Jean will perceive her maximum loss to be when John spends a lot on arms, and this will be a minimum when she does also. Perhaps this is why the minimax method is so popular with defence establishments because it involves more expenditure on defence and enhances their position.

Table 10.6

	Jean	Heads	Tails	Mix
John	Heads	$(1, -1)$	$(-1, 1)$	$(0,0)$
	Tails	$(-1, 1)$	$(1, -1)$	$(0,0)$
	Mix	$(0,0)$	$(0,0)$	$(0,0)$

Remember, it is the corporate utility that is needed, not that of any set of individuals sharing a common interest.

There are other methods of decision-making. Several of them are easily exposed as unsatisfactory. One, much favoured by statisticians, is too technical for us to discuss here but is open to the usual criticism of incoherence. So let us turn to quite a different objection, namely an objection to the principle of coherence.

10.13 COHERENCE

One objection that has been raised against what we have called coherence and what is often called consistency has been epigrammatically expressed by saying 'it is better to be inconsistent but sometimes right, than to be consistently wrong'. Before discussing this objection it is necessary to think about what is meant by a decision being right (or wrong). It is common to hear people say that a decision was ridiculous. For example, the defences of Singapore in the 1940s were designed to withstand an assault from the sea and were useless against a land-attack. In the event, the Japanese came from the land. We might say that the original decision about the defences was wrong. Let us try to analyse what is meant here. I think that what is most often meant is that some contingency which ultimately happened was not considered when the original decision was taken. (For example, the attack on Singapore from the land.) If so, this amounts to a criticism that the decision-maker forgot something. This may happen; even with the formal system it has been stressed that no scientific, systematic, logical procedure can make sure that the lists are exhaustive. All that can be said is that since the formalism requires such a list it encourages an attempt at exhaustion but the attempt may not succeed. Another possibility is that the contingency was foreseen but dismissed as extremely unlikely. (Just as the train having an accident was omitted from the mountain-pass example in section 8.12.) Alternatively expressed, this says that the probabilities were wrong. Similarly, a decision may be criticized because the utilities were incorrect.

So, aside from having forgotten something, a comment that a decision was wrong often means that some probabilities or utilities were wrong. Therefore we must enquire that we might mean by these being incorrect. Often the criticism is entirely misplaced because the objector is using information that was not available to the original decision-maker. He is using hindsight. Probabilities are conditional on the circumstances in which their assessment is being made, and on the information available then. They change, as the decision-maker's knowledge changes, according to rules that have been investigated. Similar remarks apply to the utilities. It is therefore important to remember the circumstances of the original decision-maker when criticizing his performance. Most adverse comments are misplaced for this reason.

There remain those occasions where it is felt that the original probability assessment was wrong in the light of the information available at the time. To be precise, suppose θ is an uncertain event and A describes the conditions under which its uncertainty is to be evaluated. In what sense can some value, $p(\theta \mid A)$, be correct? We have not given any such definition, though it has been suggested in the discussion on committees that provided A is well-defined most people will agree on a value. But most people were wrong over Singapore. There are some probabilities that are almost universally accepted. For example, if A includes extensive knowledge about a coin and θ is the event that it falls heads when reasonably tossed, then it would be an unusual person who came up with $p(\theta \mid A)$ anything other than 1/2. But if John insists that $p(\theta \mid A) = 1/3$ who is to say he is wrong? He will be wrong if he fails to react to data on tosses of the coin by using Bayes' theorem, as in the example of section 6.12, but I can see no sense in which his original curious value is wrong. The only way he can be wrong is in not being coherent.

Usually a person is 'wrong' over his probabilities only in the sense that they do not cohere. If John has $p(\theta \mid A) = 1/3$, one could probably investigate other beliefs that he has about chance mechanisms and demonstrate incoherence. And that is all I can see in saying that he is 'wrong'.

Consequently the reply to the epigram quoted above is that right and wrong, in the context, do not appear to have a well-defined meaning outside coherence. Incidentally, if they did and the correct decision could be recognized, one's decision-making in the other cases would presumably be improved by invoking coherence.

10.14 HINDSIGHT

It is perhaps worth enlarging on the point made in the last section about hindsight because it can have important practical consequences. We illustrate by discussing the complex situation of medical malpractice and its entanglement with the legal profession, especially in the United States.

Suppose that a doctor is uncertain what disease a patient has, one of the possibilities being cancer, which we denote by θ. He has available information H and considers $p(\theta \mid H)$, finds it to be small, and treats for some ailment other than cancer. Subsequently it is found that the patient has cancer; it is then too late to effect a cure and the patient dies. The lawyer enters and the doctor is accused of malpractice, meaning that the doctor acted wrongly *in the light of the information then available to him*. It is the phrase in italics that is important.

Malpractice could mean that $p(\theta \mid H)$ was incorrect. There is now additional evidence E from the autopsy and $p(\theta \mid H, E)$ is almost 1. (Not quite, for autopsies are occasionally wrong; remember Cromwell (section 6.7).) It is very difficult for anyone to forget E and go back to assess $p(\theta \mid H)$. One possible way is to use Bayes' theorem in reverse, and consider also $p(E \mid \theta, H)$ and $p(E \mid \bar{\theta}, H)$. That is difficult here where they are very near 1 and 0, respectively. And then there is the general point made above that eccentric probabilities cannot be condemned just because they are unusual, only because they are incoherent. So we might check the doctor's coherence, not with E but by other, possible hypothetical, evidence. This leads to a second line of attack.

This says that, knowing only H, the doctor should have carried out further tests to investigate the possibility of cancer, and the malpractice lies in his failure to do this rather than in his faulty probability $p(\theta \mid H)$. Suppose there exists a test that could have resulted in F_1 or F_2: for example, positive or negative evidence for cancer. Should it have been performed? We have seen how to answer this question in Chapter 7, where the value of information expected from the test was calculated. A legitimate medical defence would be that the value did not exceed the cost of the test on the evidence H then available. This could be checked and would involve the doctor's coherence, or lack of it. Admittedly, $p(F_i \mid \theta, H)$ and $p(F_i \mid \bar{\theta}, H)$ would need to be considered but there is often much more agreement about these than about probabilities of θ. Malpractice would be established if the evaluations were in error or if the value of information was greater than the cost. The suggestion has been made that one cannot cost a test and that a test should always be performed. This appears to happen: doctors carry out many tests, often quite incoherently, in order to cover themselves against a charge of malpractice.

There remains the possibility of an incorrect utility function. In medical questions it is surely the patient's utility that is relevant, not the doctor's. The patient would do well to listen to the doctor's advice but fundamentally the choice rests with him.

Hindsight, involving the forgetting of known evidence, is a difficult subject and it is not only lawyers but all of us who have problems in making allowance for it.

10.15 CONCLUSION

I cannot pretend to have dealt with all the substantial objections to maximization of expected utility; for example, the protest that Galsworthy puts into the mouth of one of his characters and which is quoted at the beginning of this chapter. (To judge from the *Forsyte Chronicles*, Galsworthy must have thought a good deal about decision processes in the later years of his life.) But I have considered the more important and frequently occurring ones amongst those that I have encountered. Where does the method of maximum expected utility stand at the present time?

It is as a general framework for practical decision-making that the structure developed in this book is likely to be most useful. The theory shows rather clearly what are the major elements needed in any and every problem of choice between acts. These are: the uncertain events, the decisions, the chances of the events, and the desirabilities of the consequences. It is a rewarding exercise to use this as a basis for practice even if, at some stage in the analysis, one has to omit full rigour and approximate. The theory remains as an ideal to which one aims and which must not be grossly violated without serious consequences. Without such an ideal one has no criterion for evaluating one's actions. An engineer always has to remember and use Newton's laws: he may deviate from their exact letter, he may approximate ruthlessly, but they are always there, correct and undeniable. So is the rule of maximization of expected utility.

The framework does not require anything sophisticated in the way of mathematics, though it does require a little logical abstraction. It should, therefore, be appreciated by even the most innumerate administrator. I cannot speak for other countries, but in Britain our administrators are, by and large, either self-made men who have had little education, or arts graduates who, finding little use for their philosophy, history, or languages, move to a field which, until recently, required little expertise. The former group have acquired administrative skills by experience: the latter have used their academic training to acquire some knowledge of the art. But neither group has gathered much in the way of numerate skills. Even in industry, where engineers and scientists design and operate the machines, the bosses may be totally innumerate and rely on intelligence, personality, and energy to maintain their position. Hopefully, such people could understand and use the structure for the decision-making process advocated in this book.

The situation is different when it comes to the numerical expression of the uncertainties and the desirabilities of the consequences. Two difficulties then arise: first, the quantitative expression of the ideas, second, the manipulation of these numbers. Here the sort of training (or lack of it) that these people have had will not help them to appreciate the analysis involved. The rules of probability are subtle and not all that easy to understand. There will almost certainly be considerable difficulty in implementing this part of the programme. We, in Britain, desperately need more senior people with an appreciation of and a sympathy with quantitative ideas. Until we have them we must be content with trying to impress on those responsible for important decisions at least the need to consider how likely certain contingencies are to arise and how much better some are than others. I hope that this book may have persuaded some that the methods advocated therein have a relevance to their activities. It is by a blend of the logical, scientific, numerate ideas given here, and those skills of a less formal nature, that we are best able to progress. Maximization of expected utility is an essential tool in decision-making. But not the only one.

Appendix

The tables provide two examples of utility functions for money. Table A.I is for a decision-maker with constant risk-aversion: Table A.II for someone with decreasing aversion to risk. For a sum of money equal to x dollars the tables give the corresponding utility $u(x)$, correct to 3 places of decimals. Thus for II with $x = 34$ we have a utility of 0.487 utiles for 34 dollars.

The successive values of x are at intervals of one dollar for the smallest values of x, for higher values the intervals change first to 5 and then to 10, and, for a brief section in Table A.II, to 100. For the highest values of x *critical values* are given. A critical value for x of $u(x)$ means that the utility equals that value until the next critical value is reached. Thus in Table A.I against $x = 541$ we have a utility of 0.996: this means that the utility stays at 0.996 from $x = 541$ until the next cited value, namely $x = 566$, where it reaches 0.997. Thus $u(552) = 0.996$. $u(x) = 1$ for all x greater than 761. Similar remarks apply to Table A.II.

The utility for other values of x may be obtained by *interpolation*. This can best be explained by examples. Table A.II has the successive entries:

x	$u(x)$
230	0.842
240	0.849

Thus an increase of 10 in x gives an increase of 7 in the last place of $u(x)$. Hence if we want the utility for $x = 233$, an increase of 3 in x from 230, we argue that this must give an increase of 3/10 of 7, namely 21/10, or 2, in utility. Hence $u(233) = 0.844$, an increase of 2 over 0.842.

The tables may also be used to find what sum of money has a given utility. Thus in Table A.II, a utility of $1/2(= 0.500)$ is provided by 36 dollars. Generally this procedure will require interpolation and only rarely will the given utility appear in the table. We illustrate using Table A.II to find what sum of money has a utility of 0.700. We find successive entries

x	$u(x)$
100	0.693
105	0.702

Table A.I. Utility function for a decision-maker with constant risk-aversion

x	$u(x)$ 0.	x	$u(x)$ 0.	x	$u(x)$ 0.	x	$u(x)$ 0.	x	$u(x)$ 0.	x	$u(x)$ 0.
0	000	30	259	60	451	90	593	175	826	350	970
1	010	31	267	61	457	91	597	180	835	360	973
2	020	32	274	62	462	92	601	185	843	370	975
3	030	33	281	63	467	93	605	190	850	380	978
4	039	34	288	64	473	94	609	195	858	390	980
5	049	35	295	65	478	95	613	200	865	400	982
6	058	36	302	66	483	96	617	205	871	410	983
7	068	37	309	67	488	97	621	210	878	420	985
8	077	38	316	68	493	98	625	215	884	430	986
9	086	39	323	69	498	99	628	220	889	440	988
10	095	40	330	70	503	—	—	225	895	450	989
11	104	41	336	71	508			230	900	460	990
12	113	42	343	72	513	100	632	235	905	470	991
13	122	43	349	73	518	105	650	240	909	480	992
14	131	44	356	74	523	110	667	245	914	490	993
						115	683				
15	139	45	362	75	528	120	699	—	—	Critical values	
16	148	46	369	76	532						
17	156	47	375	77	537	125	713	250	918		
18	165	48	381	78	542	130	727	260	926	500	993
19	173	49	387	79	546	135	741	270	933	504	994
						140	753	280	939	521	995
20	181	50	393	80	551	145	765	290	945	541	996
21	189	51	400	81	555					566	997
22	197	52	405	82	560	150	777	300	950		
23	205	53	411	83	564	155	788	310	955	600	998
24	213	54	417	84	568	160	798	320	959	651	999
						165	808	330	963	761	1.000
25	221	55	423	85	573	170	817	340	967		
26	229	56	429	86	577						
27	237	57	434	87	581						
28	244	58	440	88	585						
29	252	59	446	89	589						

Table A.II. Utility function for a decision-maker with decreasing risk-aversion

x	$u(x)$ 0.	x	$u(x)$ 0.	x	$u(x)$ 0.	x	$u(x)$ 0.	x	$u(x)$ 0.	x	$u(x)$ 0.
0	000	30	458	60	605	150	764	350	913	Critical	
1	027	31	466	61	608	155	769	360	917	values	
2	053	32	473	62	611	160	775	370	921		
3	077	33	480	63	614	165	781	380	925	800	991
4	101	34	487	64	617	170	786	390	929	815	992
										840	993
5	123	35	493	65	619	175	791	400	932	869	994
6	144	36	500	66	622	180	797	410	936	902	995
7	165	37	506	67	625	185	802	420	939		
8	184	38	512	68	627	190	807	430	942	943	996
9	203	39	517	69	630	195	811	440	945	993	997
										1060	998
10	221	40	523	———	———	———	———	450	947	1162	999
11	238	41	528					460	950	1382	1.000
12	255	42	533	70	633	200	816	470	952		
13	270	43	538	75	645	210	825	480	955		
14	286	44	543	80	656	220	834	490	957		
				85	666	230	842				
15	300	45	548	90	676	240	849	500	959		
16	314	46	553	95	685			510	961		
17	327	47	557			250	857	520	963		
18	340	48	561	100	693	260	864	530	965		
19	352	49	566	105	702	270	870	540	966		
				110	709	280	877				
20	364	50	570	115	717	290	883	550	968		
21	375	51	574	120	724			560	970		
22	386	52	577			300	888	570	971		
23	396	53	581	125	731	310	894	580	972		
24	406	54	585	130	738	320	899	590	974		
				135	745	330	904				
25	415	55	588	140	751	340	909	———	———		
26	425	56	592	145	757						
27	434	57	595					600	975		
28	442	58	598					700	985		
29	450	59	602					800	991		

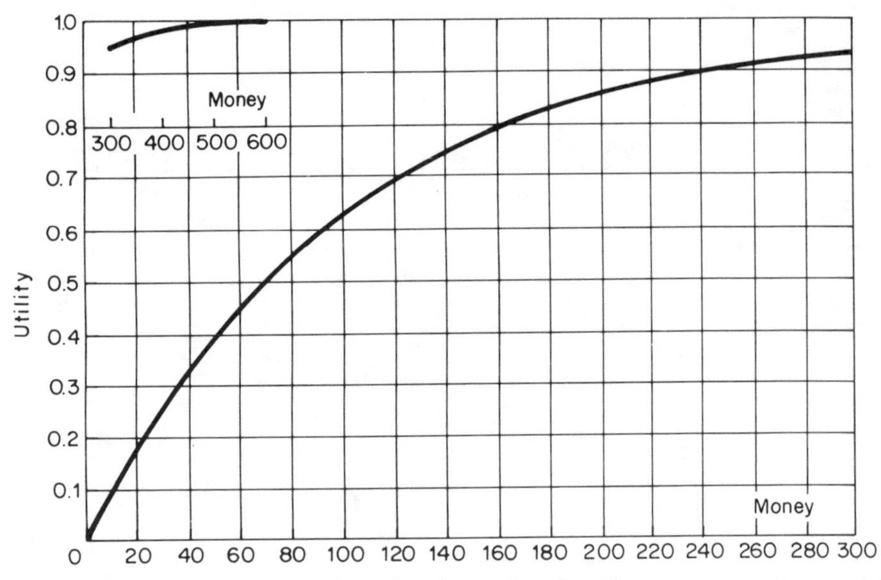

Figure A.I. Utility for money for a decision-maker, I, with constant aversion to risk

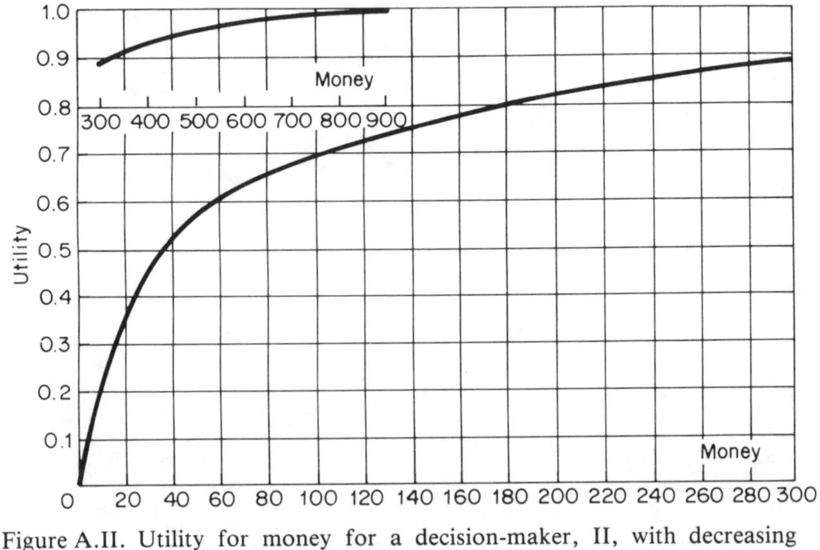

Figure A.II. Utility for money for a decision-maker, II, with decreasing aversion to risk

with utilities either side of the given value. Thus a change of 9 in the last place of $u(x)$ corresponds to a change of 5 in x. We require a change of 7 in $u(x)$ (from 0.693 to 0.700) so this must correspond to a change of 7/9 of 5, namely 35/9, or 4, in x. Hence $x = 104$ has a utility of 0.700.

Since the tables are only to three places of decimals in the utility, and to the nearest dollar, some arithmetical inaccuracies can arise in their use. Consequently readers should not worry if the values given in the text for the results of calculations, or in the answers, disagree a little (usually by 1 or 2 in the last place quoted) with those they obtain themselves.

The figures are simply graphs of the entries in the corresponding tables and may be used in place of the tables in an obvious way. Their accuracy is less than that provided by the tables. In each figure a supplementary inset figure is provided for amounts in excess of 300 dollars.

Answers to Exercises

Chapter 1

1.1. θ_1: meat and fish both good,
θ_2: meat good and fish poor,
θ_3: meat poor and fish good,
θ_4: meat and fish both poor,
d_1: order meat,
d_2: order fish.

In enlarged form there are 8 uncertain events; each of those above coupled with the wife providing meat, and each coupled with her providing fish. There is an additional decision d_3: eat elsewhere.

In the final part the 4 events that involve the meat being good may be combined to form a single event: the meat is good.

1.2. No, the exhaustive list is;
d_1: 6 5-ton lorries,
d_2: 4 5-ton, 1 10-ton,
d_3: 2 5-ton, 2 10-ton,
d_4: 3 10-ton.

1.3. Decisions: all A, all B, all C, AAB, AAC, BBA, BBC, CCA, CCB, ABC. ($m = 10$.)

Uncertain events: none, a, b, c, ab, bc, ca, abc ($n = 8$) where, for example, ab means a demand for both A and B, but not C.

If, for example, he selects AAB he will be able to satisfy all the demands: whereas ABC will fail to meet the demands associated with all the uncertain events containing a.

1.4. t_1t_2, t_1t_3, t_2t_1, $t_2t_3t_1$, t_3t_1, $t_3t_2t_1$.

Chapter 2

(T means that the unbracketed version is true: F that it is false and the one in parentheses is correct.)

Table 2.1 (1) F, (2) T, (3) T, (4) T, (5) F, (6) T, (7) F, (8) F, (9) F, (10) T.
Table 2.3 (11) T, (12) F, (13) T, (14) F, (15) F, (16) T, (17) F, (18) T, (19) T, (20) F.

Chapter 3

3.1. Probability of being in dinner jacket: 1/4.
Probability of finding the diary: 1/2 (this being the most likely of the four places).
Probability of being in left-hand pocket: 5/8.

3.2. g.

3.3. Probability that operator is idle: 0.20.
Probability at least one other stopped: 1/2.

3.4. $p(B \mid E) = 19/128$,
$p(B \mid \bar{E}) = 25/32$,
$p(B) \quad = 11/40$.

Chapter 4

4.1. (i) 0.1, (ii) 4, (iii) 3.95.
4.2. (i) $d_1(0.49)$, (ii) $d_2(0.6)$, (iii) $d_1(0.62)$.
4.3. Whether θ_1 or θ_2 is true, d_2 is better than d_1, so select d_2.
4.4 $p = 2/3$. If $p < 1/2$, then d_1.
4.5 $u = 0.1$, d_2: $u = 0.9$, d_1.
4.6. For ace or king draw a second card, for a jack decline. The expected gains are 1/8, 2/8, and 0, respectively. A fair fee is 1/9.
4.7. P must lie between 5/9 and 7/11.
4.8. $t_3t_2t_1$.

Chapter 5

5.1. Assets 30 dollars: d_2, $\bar{u}(d_2) = 0.455$,
Assets 200 dollars: d_2, since utility is there almost the same as money.
5.2. II: 14%, I: 3%, II (200 dollars): none.
5.3. I: 2%, II: none.
5.4. Assets 50 dollars: premium 3.60 dollars.
Assets 100 dollars: premium 3.00 dollars.
5.5. Assets 40 dollars: −11%.
Assets 100 dollars: +12%.
5.6. Against 40M risk: 4.5M dollars.
Against 20M risk: 1.8M dollars (approx).
5.7. II: Purchase price 8.50 dollars.
Sale price 8.75 dollars.
I: Both prices 9.4 dollars.
5.8. No gamble 0.17 utiles, one gamble 0.1696, two gambles 0.1703.

Chapter 6

6.1. $\dfrac{p_1 p(E)}{p_2 + p(E).(p_1 - p_2)}$; (a) about 1 in 10 thousand: (b) positive response, 2/3, negative response, 1/4.
6.2. (i) $0.2 - 0.12 \times P$,
(ii) $0.08 \times P/(0.2 - 0.12 \times P)$,
(iii) $0.8 - 0.6 \times P$.
(The reader can try inserting different numerical values for P.)
6.3. 16/5.
6.4. $mp/\{1 + (m-1)p\}$.
6.5. 2/3.
6.6. As it stands very little. Let D denote the description of the couple and let p_n be the probability that there are n couples in the community satisfying D, given that we already know that there is one such. Then if $n = 1$ the couple are guilty: if $n = 2$ the couple have an even chance of being guilty, and so on. For general n the chance is n^{-1}. Consequently, extending the conversation from guilt to include n, the chance of guilt is

$$\sum_{n \geqslant 1} p_n/n$$

If N is the total number of couples in the community and P is the quoted probability, this is about $1 - NP/4$, for NP small. Thus if $N =$ one million, $NP = 1/2$ and the probability of guilt is about 47/48.

6.7.

ϕ or ψ	Posterior ψ	ϕ
0.0	0.00	0.00
.1	.00	.00
.2	.00	.00
.3	.00	.00
.4	.00	.00
.5	.01	.56
.6	.04	.02
.7	.08	.03
.8	.15	.07
.9	.27	.12
1.0	0.45	0.20

Chapter 7

7.1. (i) 10, (ii) 80 (but see 7.2), (iii) 60.
7.2. 83 dollars, approximately. A rather higher figure than intuition might suggest.
7.3. No: even perfect information is worth only 3000 dollars.
7.4. $u(C+a-f)=u(C)$ for any p: hence $a-f=0$.
7.5. (i) 10/3 dollars.
(ii) Nothing, since whatever the advice is it suggests d_1.
(iii) $12\frac{1}{2}$ dollars.
7.6. 750 dollars.

Chapter 8

8.3. $c=6$ units. If $c=1$, drill if result is good or fair, otherwise do not drill. Expected profit is 19 units. (Note: the chance of getting a high yield when no seismic test has been performed must be found by extending the conversation to include the result of the test, had it been carried out. Thus $p(\text{high}) = p(\text{high} \mid \text{'good'})p(\text{'good'}) + \ldots + \ldots$ and similarly for moderate and none). Furthermore $p(\text{'good'}) = 40/130$, not 40/100, remembering the 30 additional sites where no drilling was done. These considerations are an essential part of the coherence.)
8.4. Maximum cost of test, $L/32$. For cost less than this rework only if test reveals a defect. For cost greater than this do not rework.
8.5. Operate if the chance of the patient having the disease exceeds 4/13.
8.6. Operate if the chance of the patient having the disease exceeds $4/(14-\lambda)$ where 0 is the utility of dying, 1 that of complete recovery, and λ that of partial recovery. (How does the doctor—or patient—assess the value of λ?)

Glossary of Symbols

'and'	intersection (of events)	35
b	number of black balls in the standard urn	18
C	capital, or total assets	72
C_{ij}	consequences that will result if d_i is selected and θ_j occurs	53
d_i	decision number i in the exhaustive list of exclusive decisions available to the decision-maker	6
(d_i, θ_j)	alternative form of C_{ij}	53
E	an event	18
$\bar{E}$	negation of an event E: the event that occurs if E does not occur	28, 33
E_1 and E_2	intersection (conjunction) of events E_1, E_2: the event that occurs if both E_1 and E_2 occur	35
E_1 or E_2	union (disjunction) of events E_1, E_2: the event that occurs if either E_1 or E_2 occurs	31
f	fee (for information)	119
G	gain in assets	173
H	background information (history)	21
	its omission	32
L	loss in assets	173
l_{ij}	loss that will result if d_i is selected and θ_j occurs	122
m	number of exclusive and exhaustive decisions	6
m	permium (for insurance)	28, 87
$\max_i$	maximum amongst a set of numbers labelled by the index i	58
n	number of exclusive and exhaustive events	10
N	number of balls in the standard urn	32
$O^*(E)$	odds against event E (H understood)	164
$O(E \mid H)$	odds (on) event E given information H	45
'or'	union (of events)	31
p	abbreviation for probability	22
P	abbreviation for probability	84, 162
p^2	p squared: $p \times p$	62

$p(E)$	probability of event E (H understood)	32
$p(E \mid H)$	probability of event E given information H	21
p_j	probability of event indexed by j	63, 119
pr	abbreviation for probability	95
$p(\theta_j)$	probability of event θ_j (H understood)	33
$p(\theta_j \mid H)$	probability of event θ_j given information H	113
$\bar{u}$	expected utility of the best decision	58
$u(C_{ij})$	utility of consequence C_{ij}	56
$\bar{u}(d_i)$	expected utility of decision d_i	58
u_{ij}	abbreviation for $u(C_{ij})$	119
$u(x)$	utility of money assets x	72
ϕ_j	Greek 'phi': alternative for θ_j	116
ψ_j	Greek 'psi': alternative for θ_j	112
θ_j	Greek 'theta': uncertain event indexed by j in the list of exclusive and exhaustive events	8
$\propto$	sign of proportionality	99
$\ldots$	indicating omitted elements	6
$\sum_{i=1}^{n}$	Greek 'sigma': the sum of the expressions labelled by i from 1 up to n which follow	58
,	alternative for 'and'	167
$\surd$	square root	163
$\mid$	given	21

Index

Where there are several references to a topic, a principal one is shown in bold)